Springer-Verlag Berlin Heidelberg GmbH

K. Omasa · K. Kai · H. Taoda
Z. Uchijima · M. Yoshino (Eds.)

Climate Change and Plants in East Asia

With 122 Figures, 61 in Color

Springer

Kenji Omasa
Head of Environmental Plant Science, National Institute for Environmental Studies, 16-2 Onogawa, Tsukuba, Ibaraki, 305 Japan

Keiko Kai
Senior Researcher, Global Environment Division, National Institute for Environmental Studies, 16-2 Onogawa, Tsukuba, Ibaraki, 305 Japan

Hiroshi Taoda
Section Director, Plant Ecology Section, Forest Environmental Division, Forestry and Forest Products Research Institute, Matsuno-sato, Kukizaki, P.O. Box 16, Tsukuba Norin, Ibaraki, 305 Japan

Zenbei Uchijima
Professor, School of Humanities, Miyazaki Municipal University, 1-2-2 Funatsuka, Miyazaki, 880 Japan

Masatoshi Yoshino
Professor, Institute of Geography, Aichi University, 1-1 Machihata-cho, Toyohashi, Aichi, 441 Japan

ISBN 978-4-431-66901-2 ISBN 978-4-431-66899-2 (eBook)
DOI 10.1007/978-4-431-66899-2

Printed on acid-free paper

Originally published by Springer-Verlag Tokyo 1996
Softcover reprint of the hardcover 1st edition 1996

Preface

The concentration of greenhouse gases in the atmosphere is increasing at a surprising rate mainly due to the massive consumption of fossil fuels and large-scale deforestation. This increase in greenhouse gas concentration has been considered likely to lead to a warming of the atmosphere near the Earth's surface. The first assessment of Working Group I of the International Panel on Climate Change (IPCC) presented in 1990 a report indicating a probable increase in global mean surface temperature of about 1℃ by 2025 and of about 3℃ by the end of the next century. Recently, the IPCC has worked on a new scientific assessment in which the global mean surface temperature increases at a rate of between 0.15° and 0.33℃ per decade when the effects of greenhouse gases alone are considered. If the possible (and more uncertain) effects of aerosols on the global climate are also taken into account, the projected rates of warming over the next century would be in the range 0.12° to 0.26℃ per decade. It has been stated by many scientists that such a rate of global warming may exceed the ability of many species to adapt or disperse into more favorable regions, and thus many plant and animal species may become extinct. Climate change may also affect natural vegetation and the productivity of agriculture.

The information presented in this book is the result of the research project Evaluation of the Global Warming Effects on Plants, part of the Global Environment Research Program supported by the Japan Environment Agency. In this project we have investigated the effects of climate change on plants in East Asia, which is one of the most important regions in relation to global environmental problems because of rapid population growth and active industrialization. The probable changes in climate and the modeling and prediction of plant production, vegetation distribution, phenological events, and lake ecosystems are discussed in this book. Also, the importance of analyzing local field data on microclimate, soil environment, snow patches, and flora and plant growth in mountain vegetation is shown. The effects of increasing concentrations of greenhouse gases and rising temperatures on rice yield are described from experiments using controlled environment facilities. We hope that the information presented here will make a significant contribution to the efforts of investigators engaged in the study of global warming problems and in planning the research strategies to be followed.

We would like to acknowledge the support of Dr. Shoji Kusunoki and other members of the Global Environmental Division of the Japan Environment Agency and Drs. Masayuki Yasuno, Shuzo Nishioka and Tsuneyuki Morita of the National Institute for Environmental Studies. We also thank Mr. Kazuo Tobe, who assisted in editing the book, and the many researchers and assistants who were of great help in concluding these studies.

Kenji Omasa
Keiko Kai
Hiroshi Taoda
Zenbei Uchijima
Masatoshi Yoshino

Contents

II. Mountain Vegetation

Contributors

CHIBA, YUKIHIRO, Forestry and Forest Products Research Institute, P.O. Box 16, Tsukuba Norin, Ibaraki 305, Japan.

DAIMARU, HIROMU, Tohoku Research Center, Forestry and Forest Products Research Institute, 72 Nabeyashiki Shimokuriyagawa, Morioka, Iwate 020-01, Japan.

FUJIMOTO, KIYOSHI, Forestry and Forest Products Research Institute, P.O. Box 16, Tsukuba Norin, Ibaraki 305, Japan.

GOTO, YOSHIAKI, Kansai Research Center, Forestry and Forest Products Research Institute, Kyoto, 612, Japan.

HORIE, TAKESHI, Faculty of Agriculture, Kyoto University, Kyoto 606, Japan.

HOSOMI, MASAAKI, Faculty of Engineering, Tokyo University of Agriculture and Technology, Koganei, Tokyo 184, Japan.

IKEDA, SHIGETO, Tohoku Research Center, Forestry and Forest Products Research Institute, 72 Nabeyashiki Shimokuriyagawa, Morioka, Iwate 020-01, Japan.

IKEGUCHI, HITOSHI, Museum of Nature and Human Activities, Sanda, Hyogo, 669-13, Japan.

ISHIZUKA, MORIYOSHI, Forestry and Forest Products Research Institute, P.O. Box 16, Tsukuba Norin, Ibaraki 305, Japan.

KAI, KEIKO, National Institute for Environmental Studies, Tsukuba, Ibaraki 305, Japan.

KAINUMA, MIKIKO, National Institute for Environmental Studies, Tsukuba, Ibaraki 305, Japan.

KAJIMOTO, TAKUYA, Tohoku Research Center, Forestry and Forest Products Research Institute, 72 Nabeyashiki Shimokuriyagawa, Morioka, Iwate 020-01, Japan.

KOBAYASHI, SHIGEO, Forestry and Forest Products Research Institute, P.O. Box 16, Tsukuba Norin, Ibaraki 305, Japan.

KURACHI, NAHOKO, Hokkaido Research Center, Forestry and Forest Products Research Institute, Hitsujigaoka-7, Toyohira-ku, Sapporo, Hokkaido 062, Japan.

LIU, CHUNYING, Institute of Botany, Chinese Academy of Sciences, 141 Xizhimenwai Avenue, Beijing, P.R. China.

MATSUI, TSUTOMU, Faculty of Agriculture, Kyoto University, Kyoto 606, Japan.

MORISAWA, TAKESHI, Kiso Experimental Station, Forestry and Forest Products Research Institute, Kisofukushima, Nagano 397, Japan.

MURAKAMI, AKIHIKO, Faculty of Engineering, Tokyo University of Agriculture and Technology, Koganei, Tokyo 184, Japan.

MURAKOSHI, NAOMI, Faculty of Science, Shinshu University, Asahi 3-1-1, Matsumoto, Nagano 390, Japan.

NAKAGAWA, HIROSHI, Faculty of Agriculture, Kyoto University, Kyoto 606, Japan.

OHBA, HIDEAKI, University museum, University of Tokyo, Hongo 7-3-1, Tokyo 113, Japan.

OHTA, SHUNJI, School of Human Sciences, Waseda University, 2-579-15, Mikajima, Tokorozawa, Saitama 359, Japan.

OHTANI, YOSHIKAZU, Forestry and Forest Products Research Institute, P.O. Box 16, Tsukuba Norin, Ibaraki 305, Japan.

OKUBO, TAKUYA, Lake Biwa Research Institute, Otsu, Shiga 520, Japan.

OMASA, KENJI, National Institute for Environmental Studies, Tsukuba, Ibaraki 305, Japan.

PARK ONO, HYE-SOOK, Faculty of Humanities and Social Science, Mie University, Tsu, Mie 514, Japan.

SAIGUSA, TETSU, Faculty of Engineering, Tokyo University of Agriculture and Technology, Koganei, Tokyo 184, Japan.

SAKAI, HISAO, Forestry and Forest Products Research Institute, P.O. Box 16, Tsukuba Norin, Ibaraki 305, Japan.

TANAKA, NAGAHARU, Forestry and Forest Products Research Institute, P.O. Box 16, Tsukuba Norin, Ibaraki 305, Japan.

TANAKA, NOBUYUKI, Forestry and Forest Products Research Institute, P.O. Box 16, Tsukuba Norin, Ibaraki 305, Japan.

TAODA, HIROSHI, Forestry and Forest Products Research Institute, P.O. Box 16, Tsukuba Norin, Ibaraki 305, Japan.

TSUNEKAWA, ATSUSHI, National Institute for Environmental Studies, Tsukuba, Ibaraki 305, Japan.

UCHIJIMA, ZENBEI, School of Humanities, Miyazaki Municipal University, 1-1-2 Funatsuka, Miyazaki 880, Japan.

UTSUGI, HAJIME, Forestry and Forest Products Research Institute, P.O. Box 16, Tsukuba Norin, Ibaraki 305, Japan.

YAMANOI, KATSUMI, Forestry and Forest Products Research Institute, P.O. Box 16, Tsukuba Norin, Ibaraki 305, Japan.

YANG, DIANAN, Institute of Botany, Chinese Academy of Sciences, 141 Xizhimenwai Avenue, Beijing, P.R. China.

YOSHINO, MASATOSHI, Faculty of Letters, Aichi University, Toyohashi, Aichi 441, Japan.

ZHANG, JIE, Institute of Botany, Chinese Academy of Sciences, 141 Xizhimenwai Avenue, Beijing, P.R. China.

ZHANG, XINSHI, Institute of Botany, Chinese Academy of Sciences, 141 Xizhimenwai Avenue, Beijing, P.R. China.

ZHOU, GUANGSHENG, Institute of Botany, Chinese Academy of Sciences, 141 Xizhimenwai Avenue, Beijing, P.R. China.

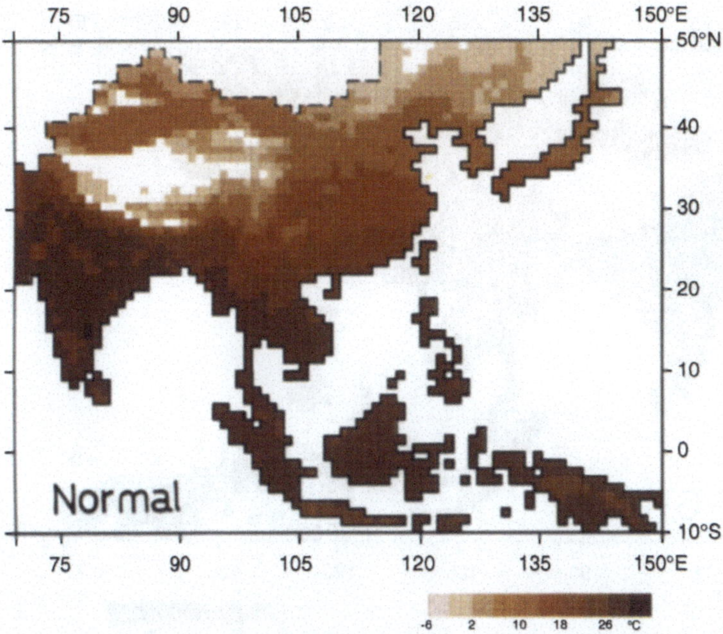

Color Plate 1a Grid mesh maps of annual mean of air temperature on the Monsoon Asia for normal and warmed climates (see p. 8). Normal.

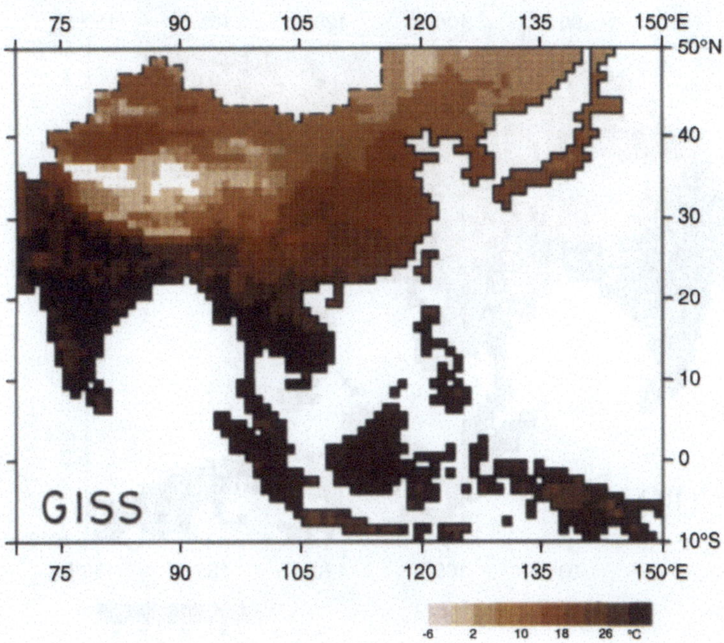

Color Plate 1b Grid mesh maps of annual mean of air temperature on the Monsoon Asia for normal and warmed climates (see p. 8). GISS-scenario.

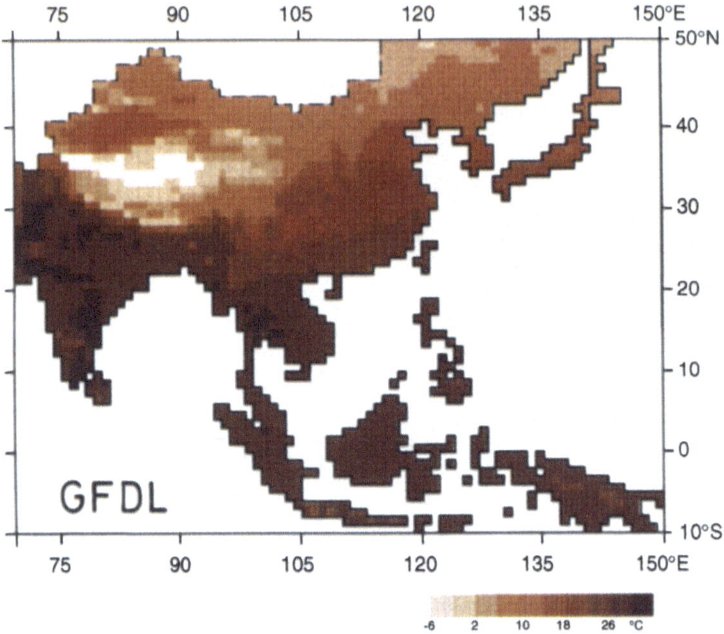

Color Plate 1c Grid mesh maps of annual mean of air temperature on the Monsoon Asia for normal and warmed climates (see p. 8). GFDL-scenario.

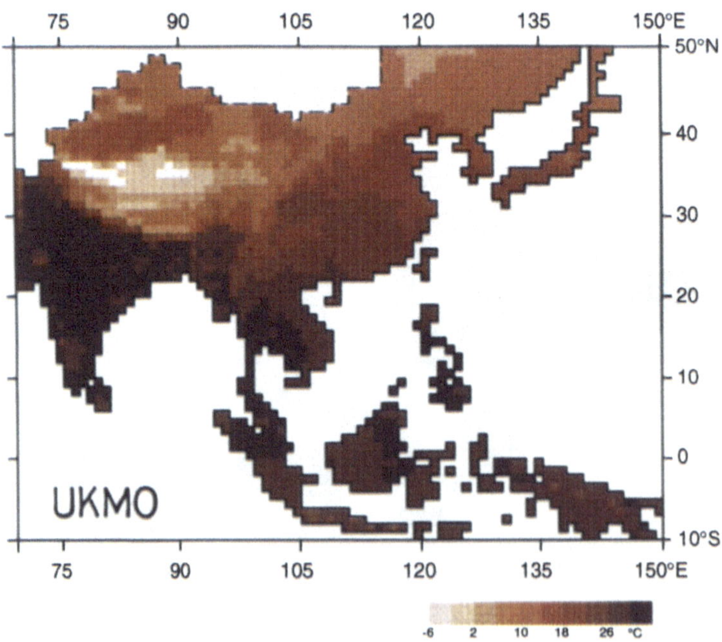

Color Plate 1d Grid mesh maps of annual mean of air temperature on the Monsoon Asia for normal and warmed climates (see p. 8). UKMO-scenario.

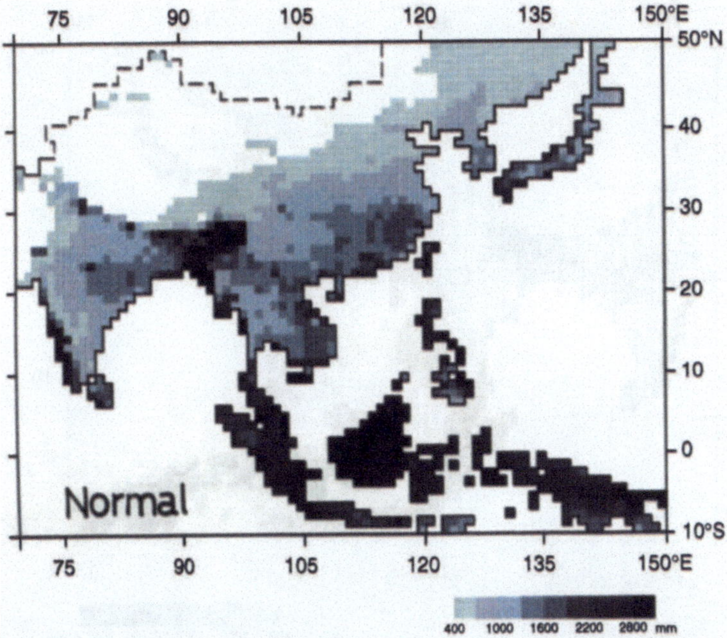

Color Plate 2a Grid mesh maps of annual precipitation on the Monsoon Asia for normal and warmed climates (see p. 8). Normal.

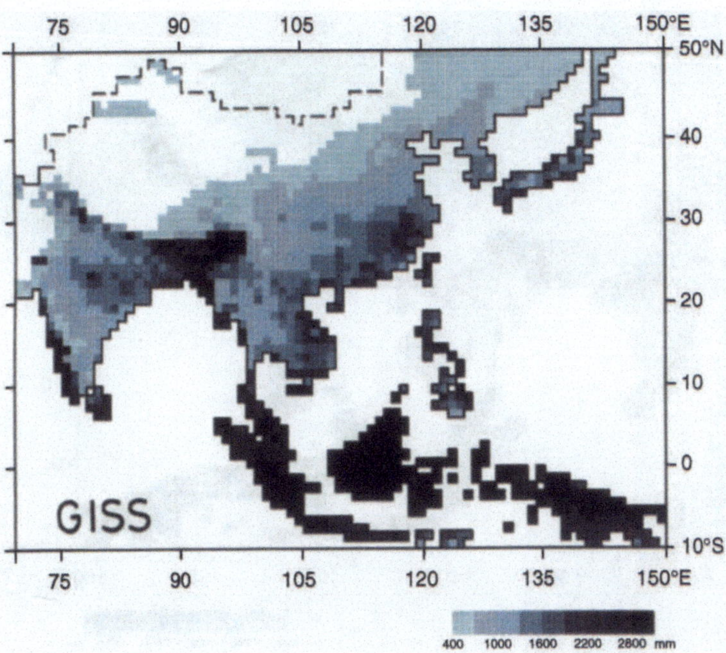

Color Plate 2b Grid mesh maps of annual precipitation on the Monsoon Asia for normal and warmed climates (see p. 8). GISS-scenario.

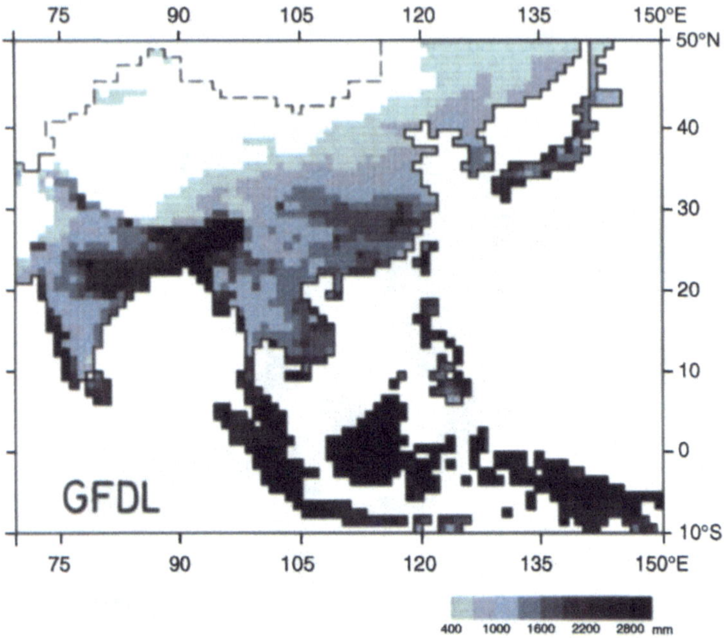

Color Plate 2c Grid mesh maps of annual precipitation on the Monsoon Asia for normal and warmed climates (see p. 8). GFDL-scenario.

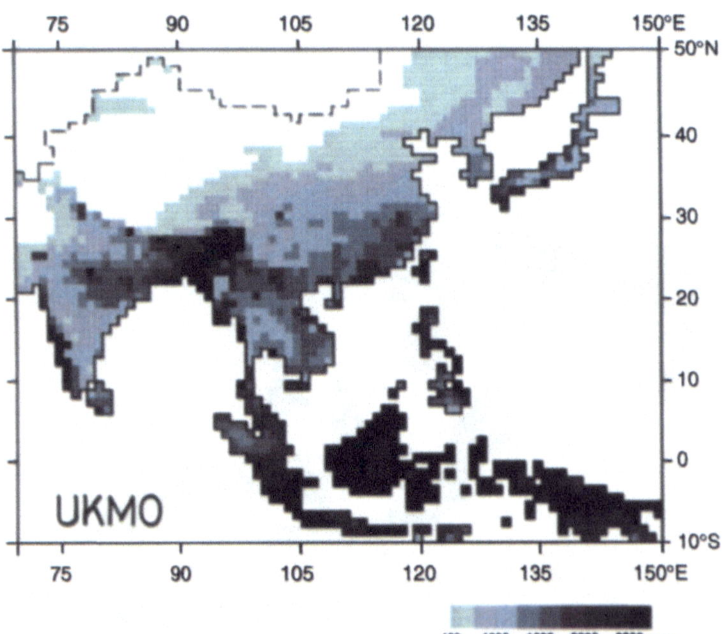

Color Plate 2d Grid mesh maps of annual precipitation on the Monsoon Asia for normal and warmed climates (see p. 8). UKMO-scenario.

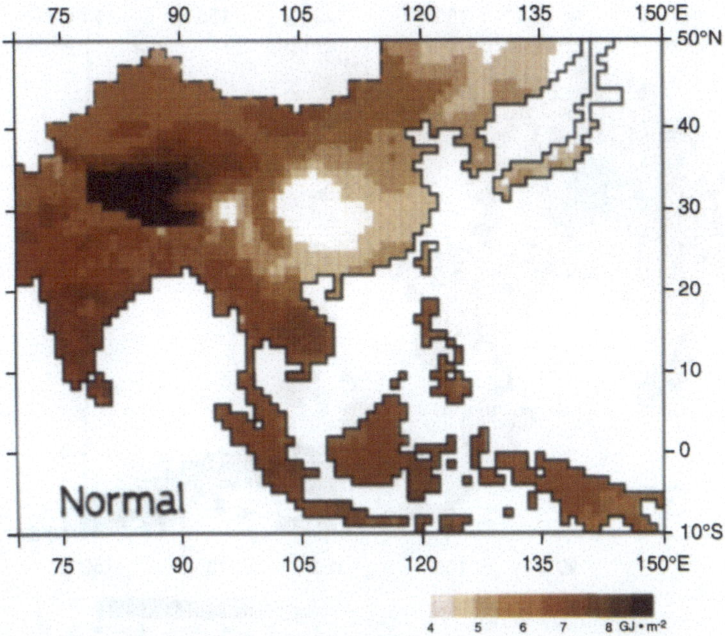

Color Plate 3a Grid mesh maps of annual global solar radiation on the Monsoon Asia for normal and warmed climates (see p. 8). Normal.

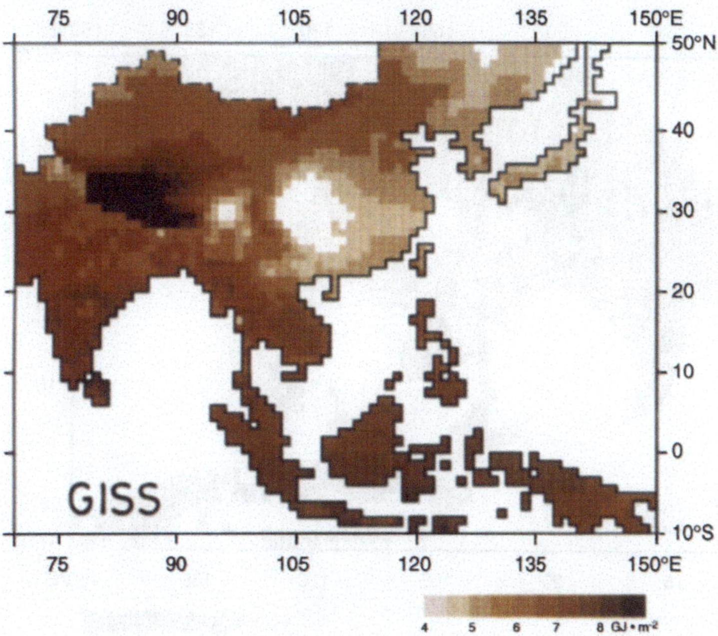

Color Plate 3b Grid mesh maps of annual global solar radiation on the Monsoon Asia for normal and warmed climates (see p. 8). GISS-scenario.

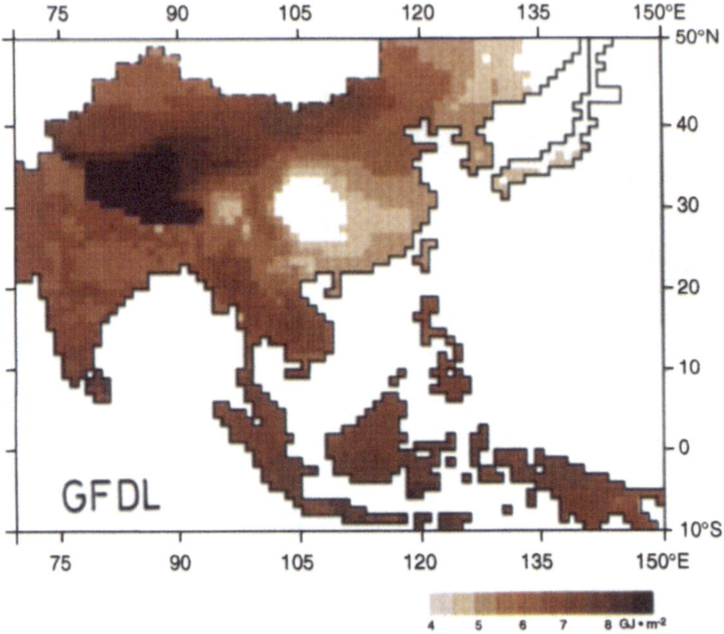

Color Plate 3c Grid mesh maps of annual global solar radiation on the Monsoon Asia for normal and warmed climates (see p. 8). GFDL-scenario.

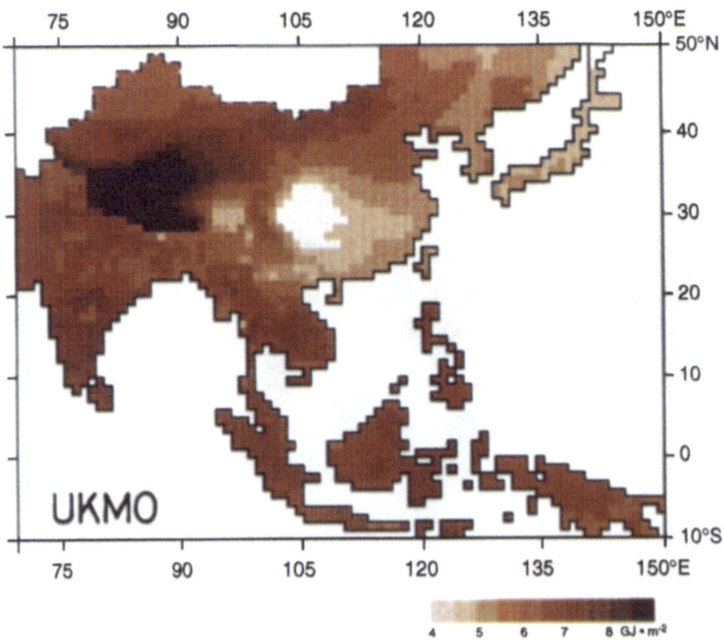

Color Plate 3d Grid mesh maps of annual global solar radiation on the Monsoon Asia for normal and warmed climates (see p. 8). UKMO-scenario.

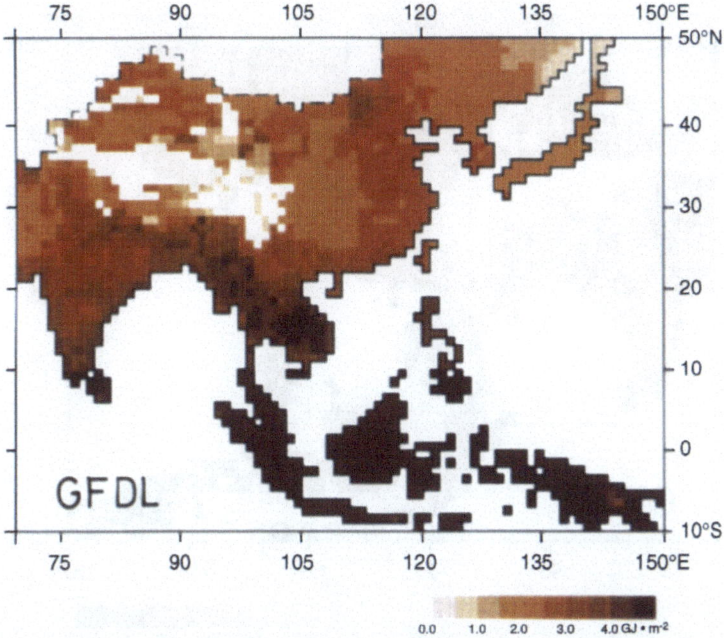

Color Plate 4a Grid mesh maps of annual net radiation on the Monsoon Asia for normal and warmed climates (see p. 9). Normal.

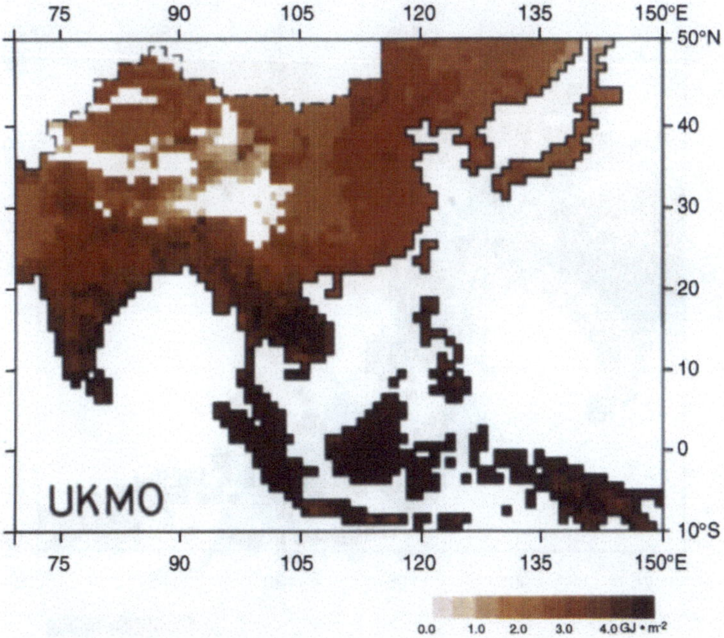

Color Plate 4b Grid mesh maps of annual net radiation on the Monsoon Asia for normal and warmed climates (see p. 9). GISS-scenario.

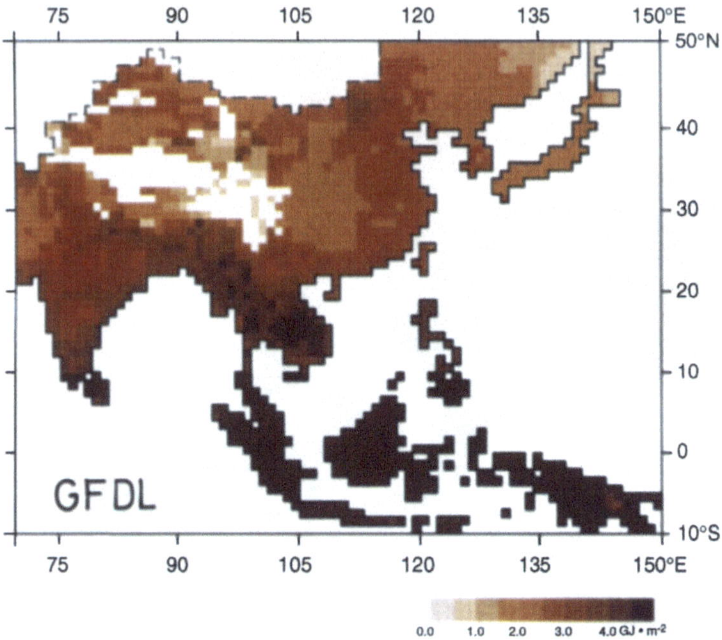

Color Plate 4c Grid mesh maps of annual net radiation on the Monsoon Asia for normal and warmed climates (see p. 9). GFDL-scenario.

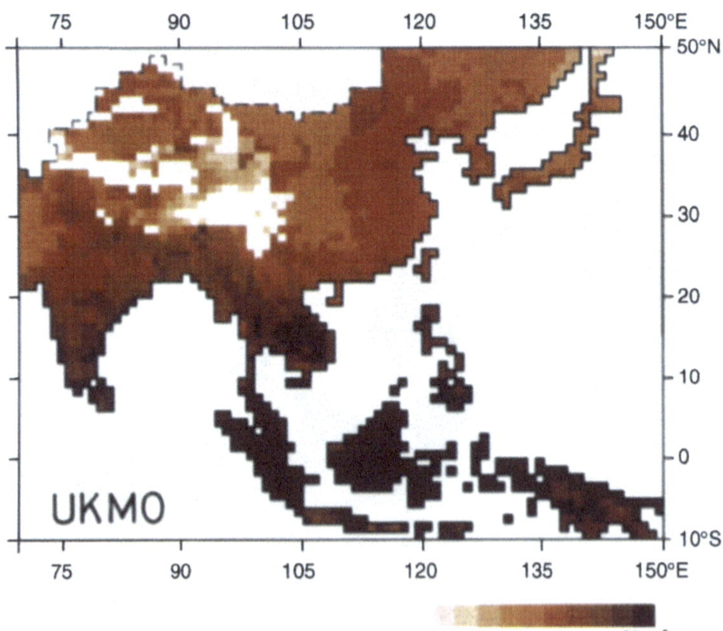

Color Plate 4d Grid mesh maps of annual net radiation on the Monsoon Asia for normal and warmed climates (see p. 9). UKMO-scenario.

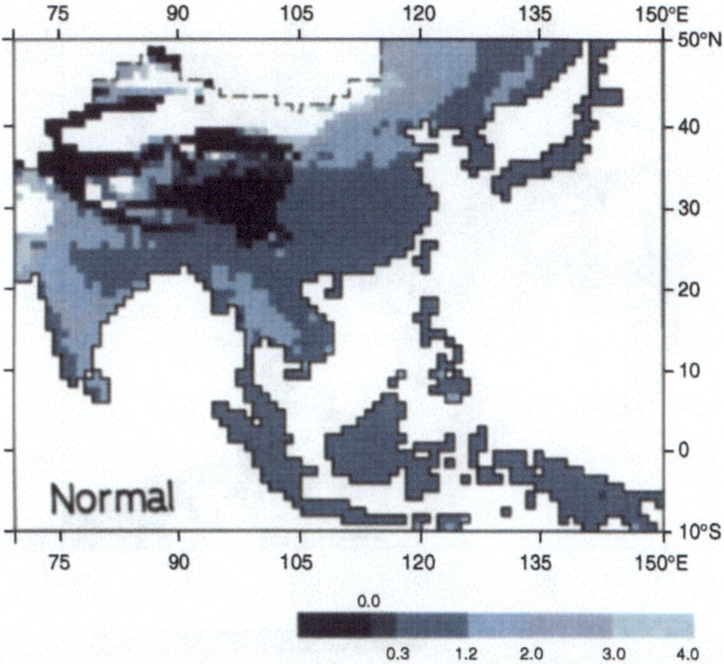

Color Plate 5a Grid mesh maps of annual mean of radiative dryness index (RDI) on the Monsoon Asia for normal and warmed climates (see p. 9). Normal.

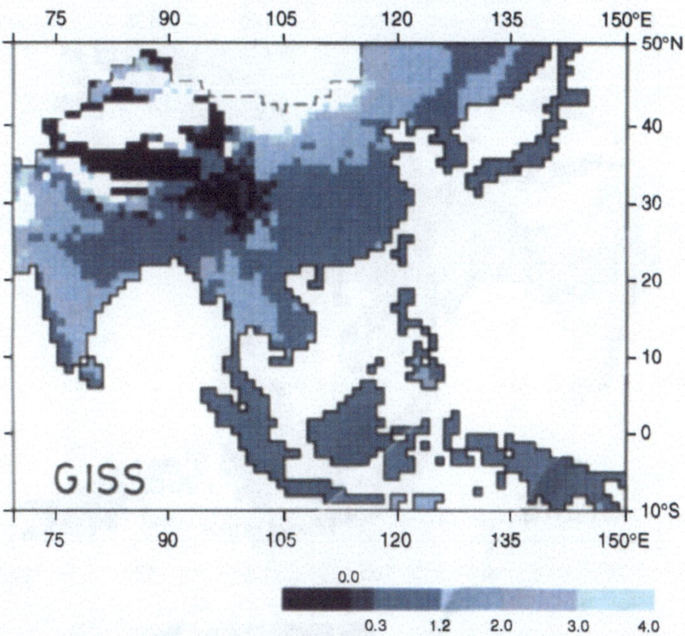

Color Plate 5b Grid mesh maps of annual mean of radiative dryness index (RDI) on the Monsoon Asia for normal and warmed climates (see p. 9). GISS-scenario.

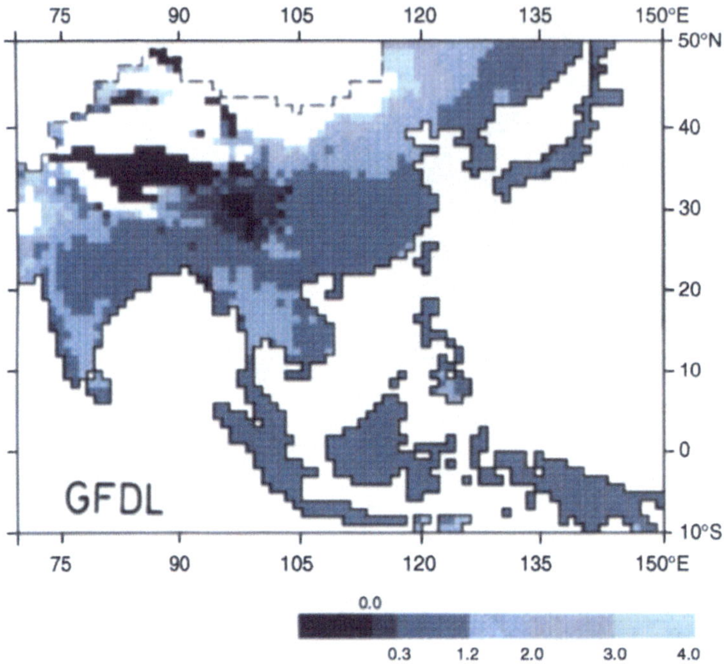

Color Plate 5c Grid mesh maps of annual mean of radiative dryness index (RDI) on the Monsoon Asia for normal and warmed climates (see p. 9). GFDL-scenario.

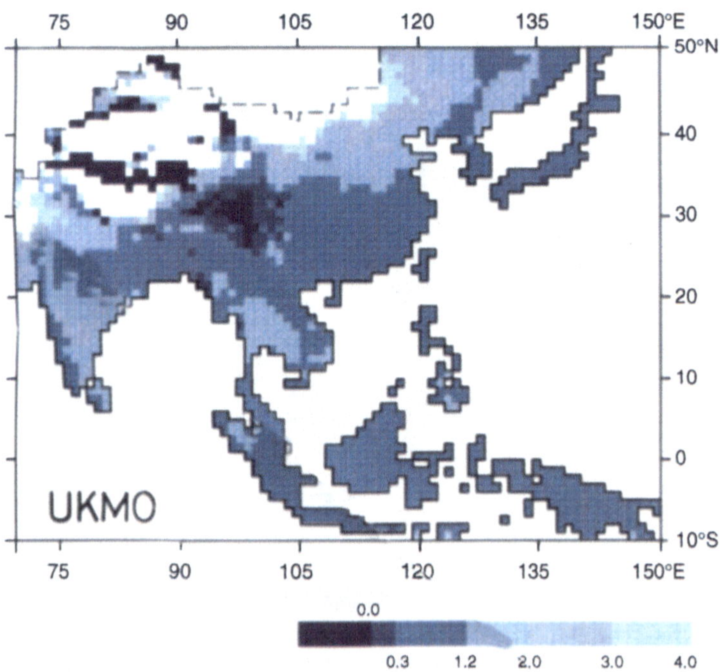

Color Plate 5d Grid mesh maps of annual mean of radiative dryness index (RDI) on the Monsoon Asia for normal and warmed climates (see p. 9). UKMO-scenario.

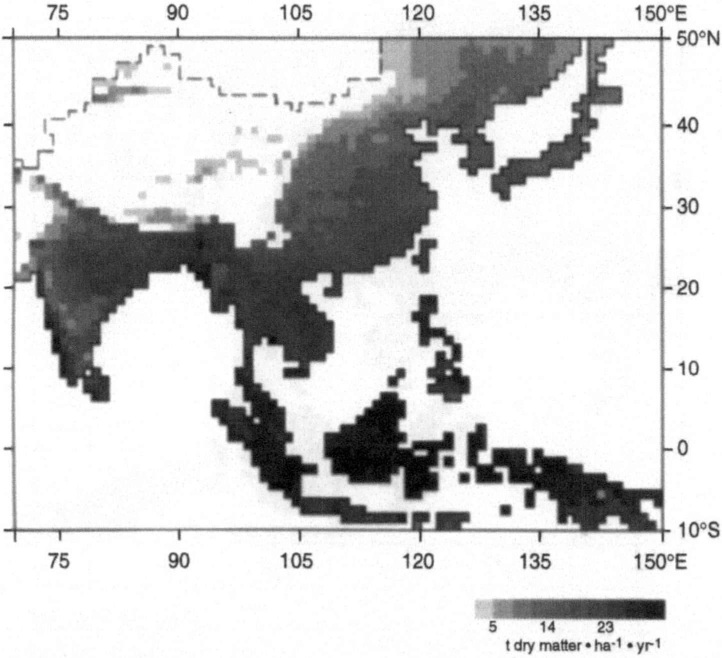

Color Plate 6 Geographical distribution of present net primary productivity (NPP) of terrestrial vegetation (see p. 17).

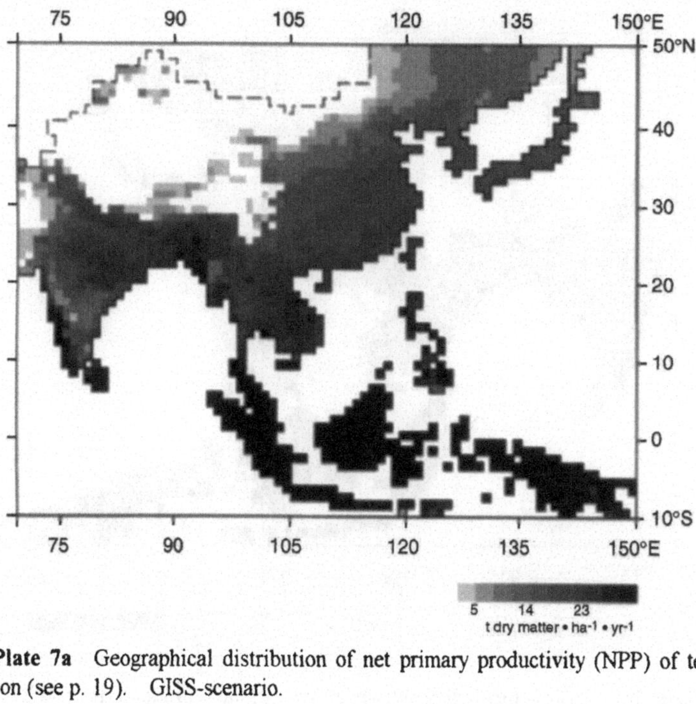

Color Plate 7a Geographical distribution of net primary productivity (NPP) of terrestrial vegetation (see p. 19). GISS-scenario.

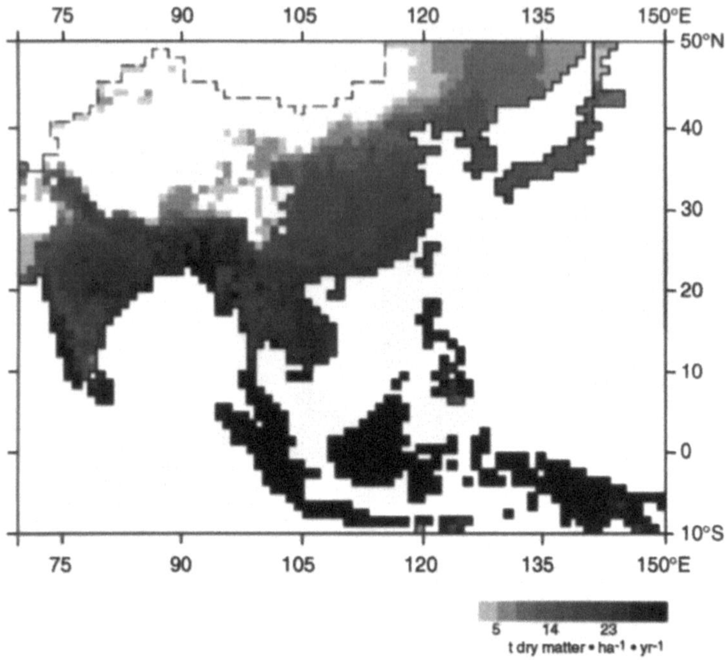

Color Plate 7b Geographical distribution of net primary productivity (NPP) of terrestrial vegetation (see p. 19). GFDL-scenario.

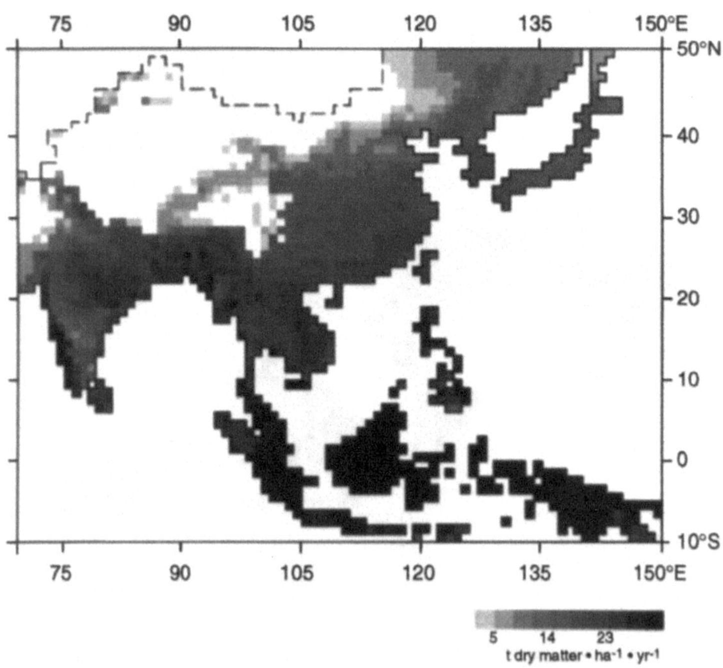

Color Plate 7c Geographical distribution of net primary productivity (NPP) of terrestrial vegetation (see p. 19). UKMO-scenario.

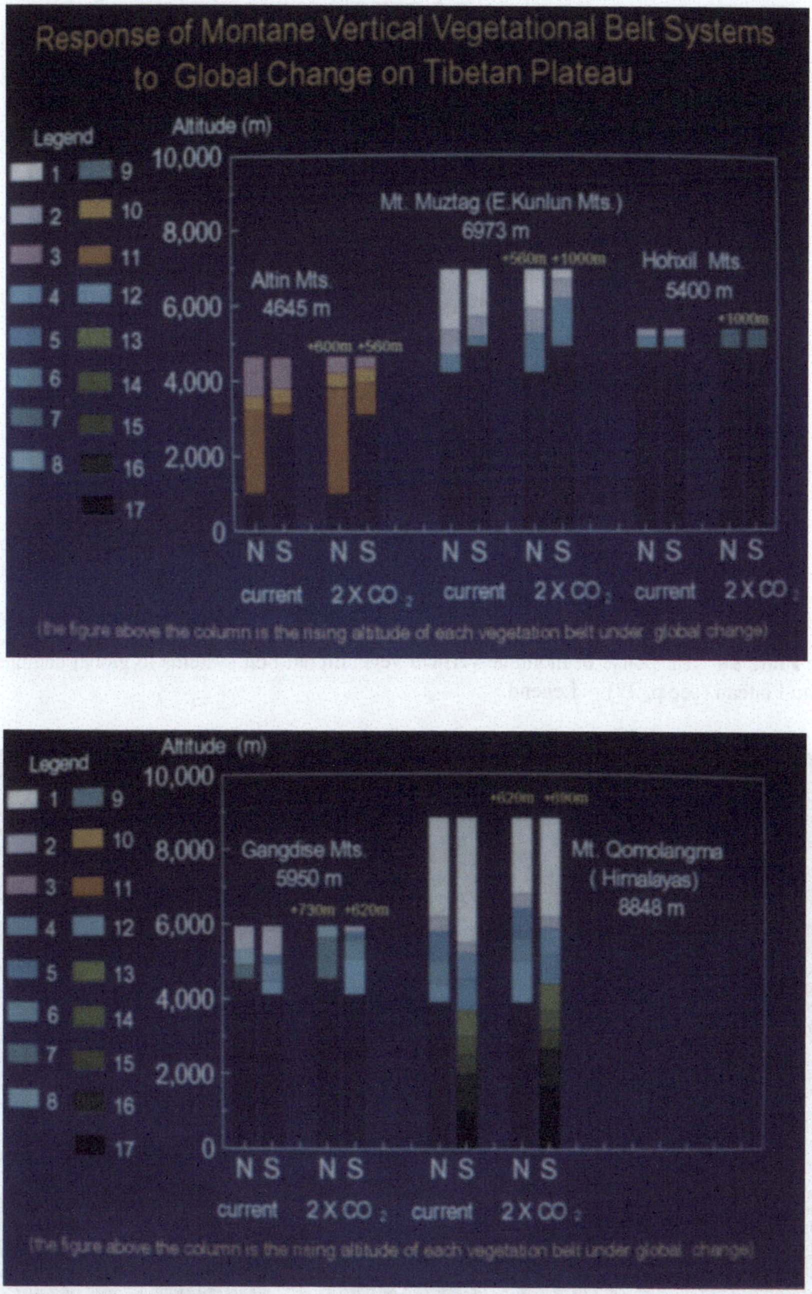

Color Plate 8a Response of montane vertical vegetational belt systems to global change on Tibetan Plateau (see p. 27).

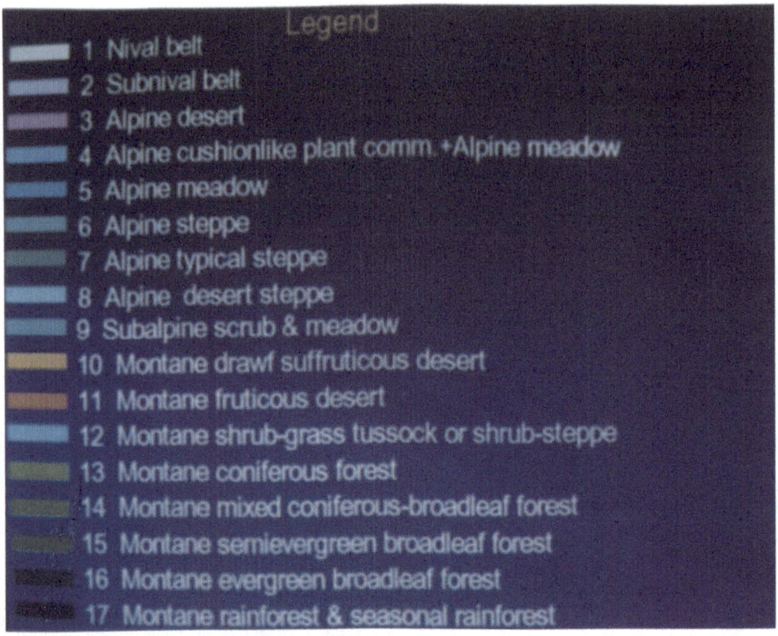

Color Plate 8b Response of montane vertical vegetational belt systems to global change on Tibetan Plateau (see p. 27). Legend.

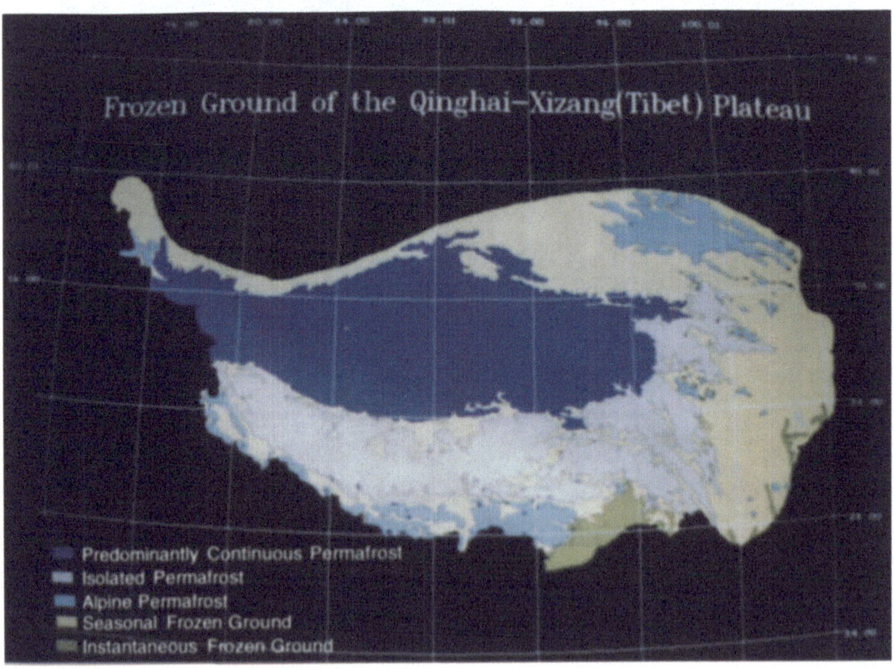

Color Plate 9 Distribution pattern of frozen ground on Tibetan Plateau (see p. 28).

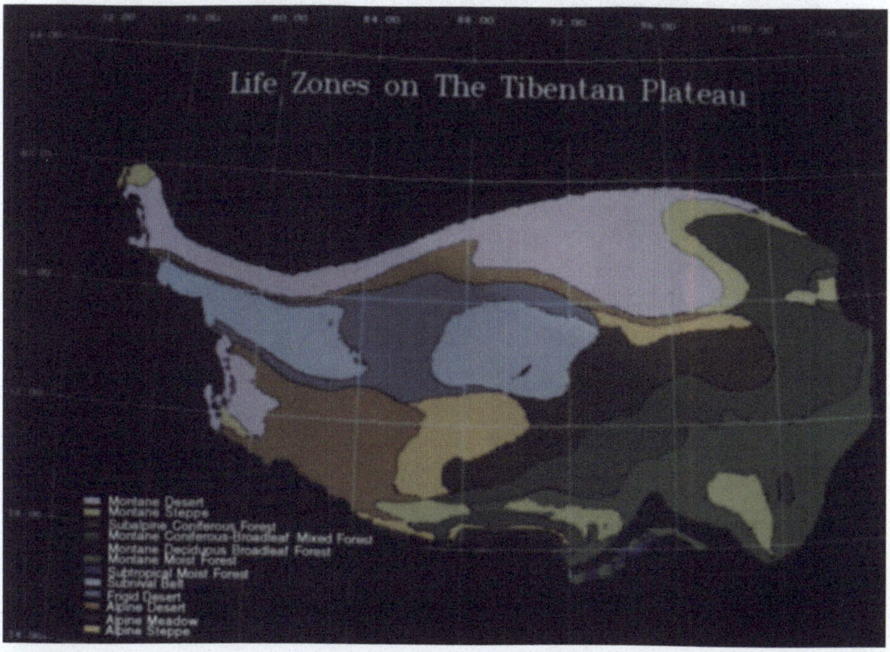

Color Plate 10 Life zones on Tibetan Plateau (see p. 29).

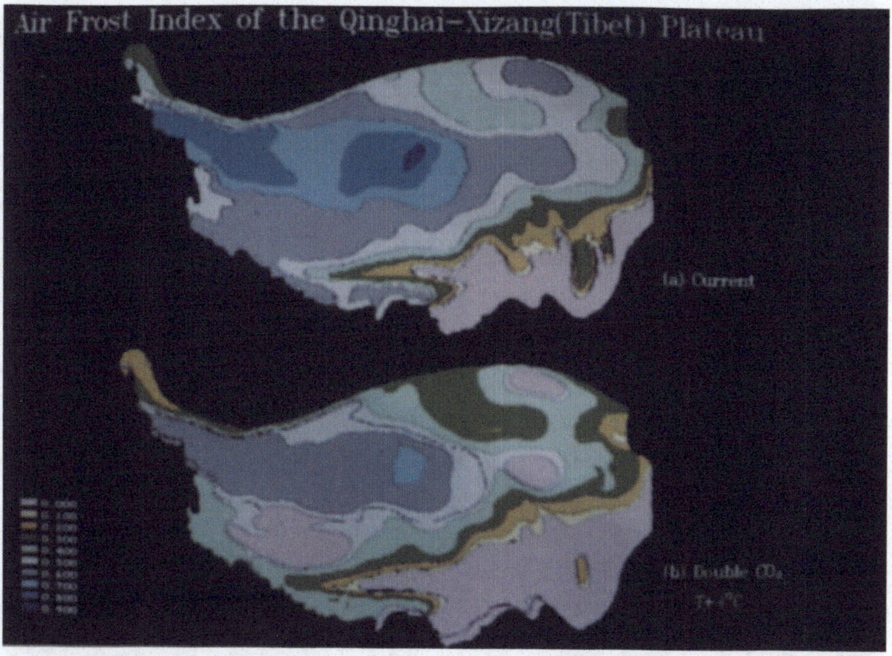

Color Plate 11 Simulation pattern and the changing pattern of frozen ground on Tibetan Plateau (see p. 30).

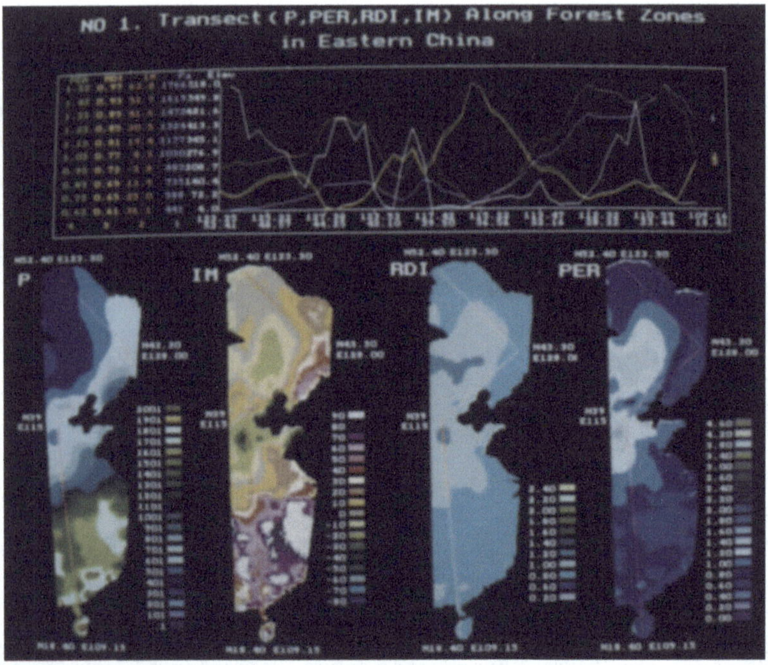

Color Plate 12 Transect (P, PER, RDI, IM) along Northeast China Transect (CENTI) (see p. 33).

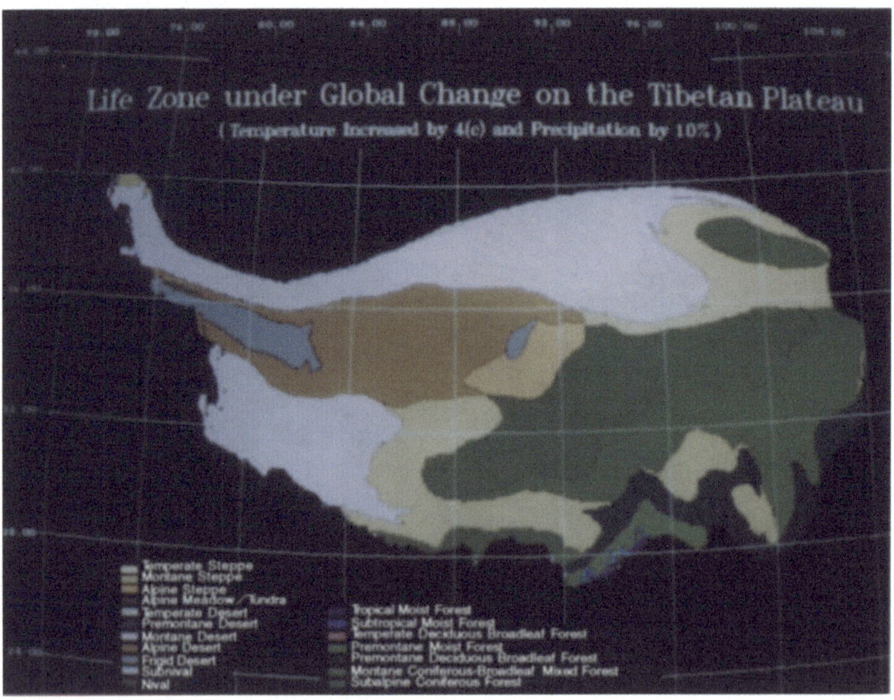

Color Plate 13 Life zones on Tibetan Plateau under global change (see p. 34).

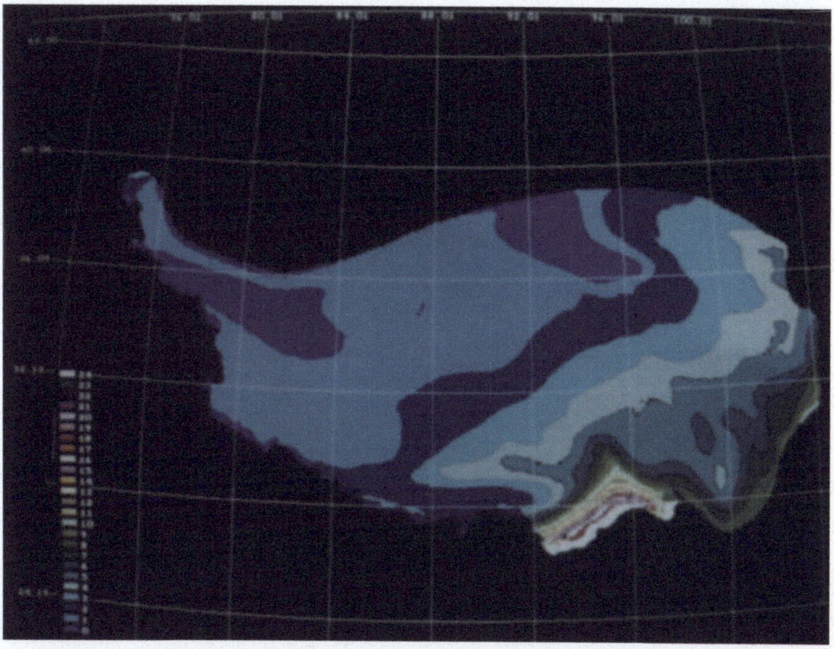

Color Plate 14 Distribution of NPP on Tibetan Plateau (see p. 36).

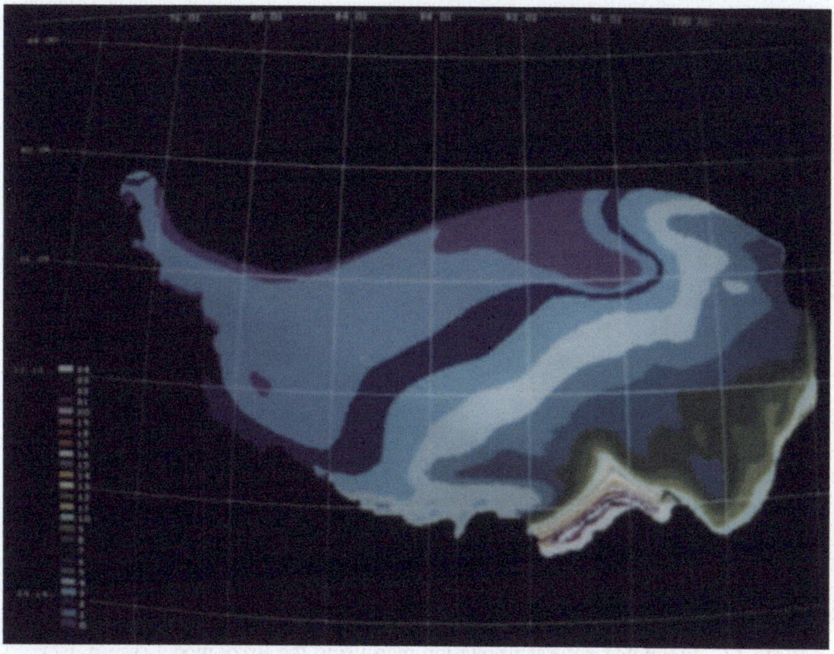

Color Plate 15 Distribution of NPP on Tibetan Plateau under global change (see p. 37).

Code	Characteristics
a	Alpine grassland, etc.
b	Subalpine conifer forest
c	Cool-temperate vegetation typical in Hokkaido
d	Cool-temperate vegetation typical in the Japan Sea side
e	Cool-temperate vegetation typical in the Pacific side
f	Warm-temperate vegetation typical on hills
g	Warm-temperate vegetation typical on lowlands
h	Subtropical vegetation
i	Substitution vegetation

Color Plate 16 Actual vegetation distribution of Japan (see p. 62).

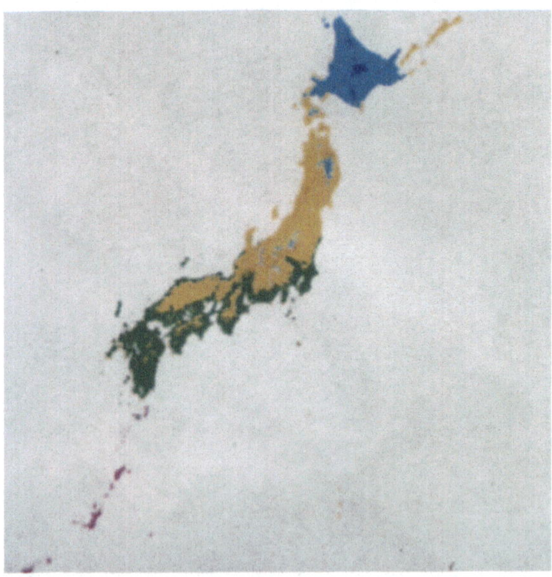

Color Plate 17 Estimated natural vegetation map using the fuzzy model (see p. 62). The legend is the same as that in Color Plate 16.

Color Plate 18 Estimated natural vegetation map using the logit model (see p. 63). The legend is the same as that in Color Plate 16.

Color Plate 19 Potential vegetation distribution predicted by the logit model under a 1-degree increase in annual mean temperature (see p. 63). The legend is the same as that in Color Plate 16.

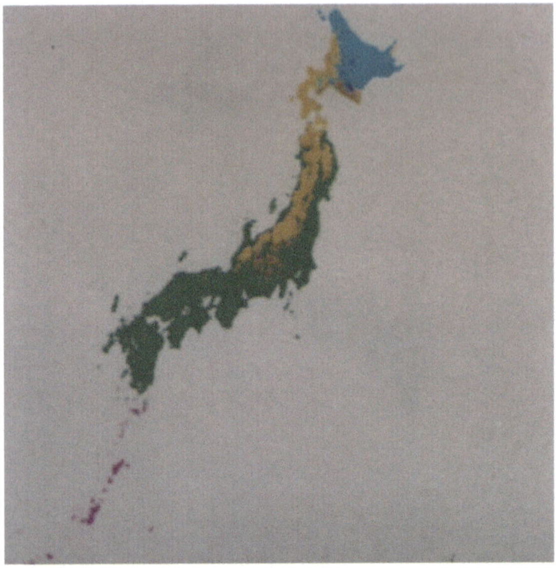

Color Plate 20 Potential vegetation distribution predicted by the fuzzy model under a 2-degree increase in annual mean temperature (see p. 63). The legend is the same as that in Color Plate 16.

Color Plate 21 Potential vegetation distribution predicted by the logit model under a 3-degree increase in annual mean temperature (see p. 63). The legend is the same as that in Color Plate 16.

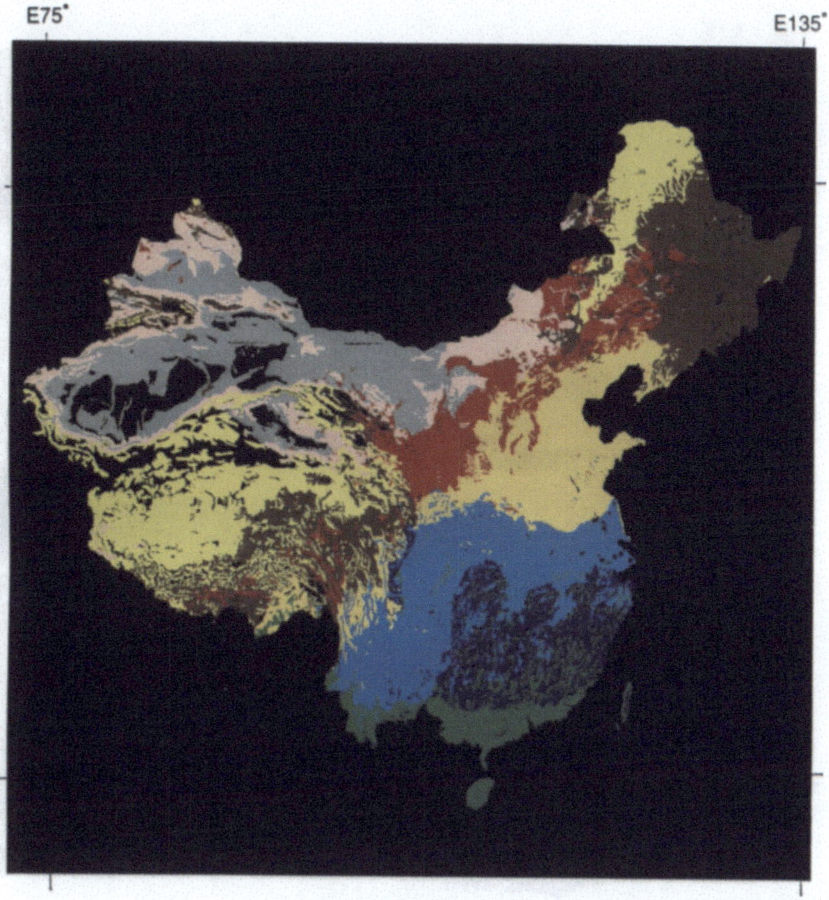

	Vegetation code	Physiognomy of vegetation
	a	Conifers/Steppes
	b	Meadows/Broadleaf deciduous forests
	c	Steppes/Deserts
	d	Steppes/Scrubs
	e	Deserts
	f	Broadleaf deciduous forests/Three crops for two years or two crops annually
	g	Evergreen and deciduous scrubs/Two crops containing upland and rice annually
	h	Broadleaf evergreen forests/Evergreen conifers/One or double cropping rice
	i	Broadleaf evergreen forests/ Double-cropping rice annually

Color Plate 22 Actual vegetation distribution of China (see p. 76). The raster-type data were digitized from the "Actual vegetation map of China (scale 1:4,000,000)" made in 1979 by the Institute of Botany, Academia Sinica.

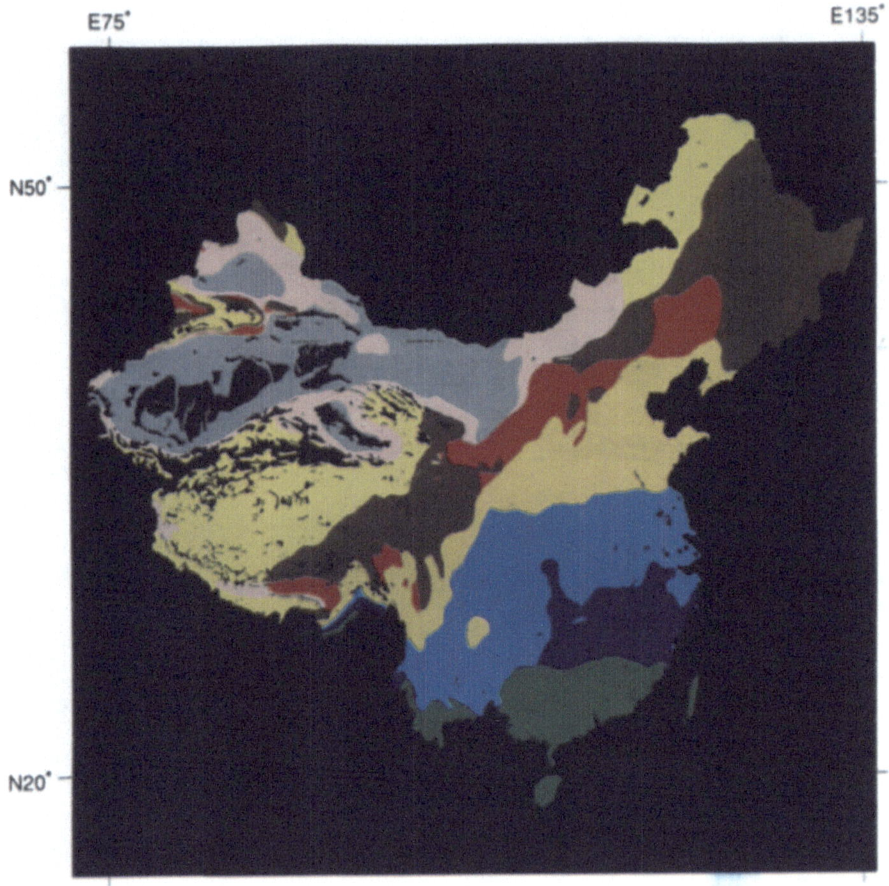

Color Plate 23 Modelled vegetation distribution (see p. 76). Here, the linear multinomial logit model was applied to explain the relationship between vegetation classification and climatic conditions. The legend is the same as that in Color Plate 22.

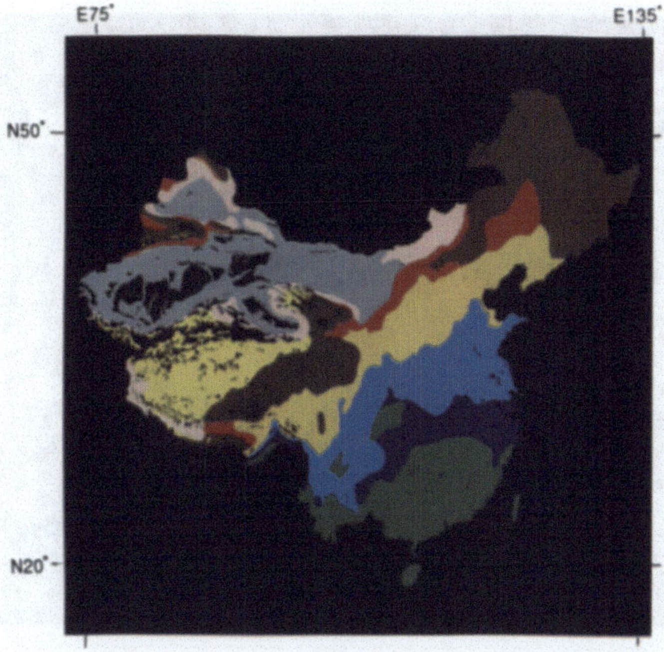

Color Plate 24 Potential vegetation distribution predicted by the logit model under Scenario-A involving a 2-degree increase in annual mean temperature and a 20% increase in annual precipitation (see p. 77). The legend is the same as that in Color Plate 22.

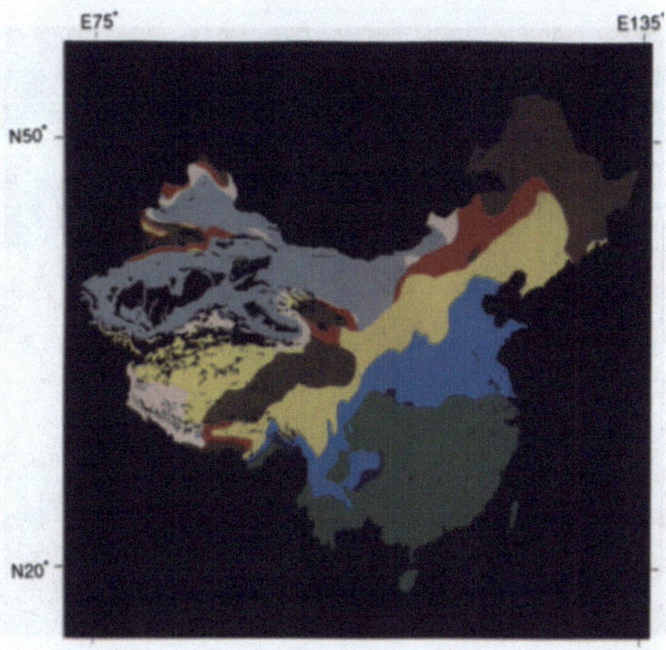

Color Plate 25 Potential vegetation distribution predicted by the logit model under Scenario-B involving a 4-degree increase in annual mean temperature and a 20% increase in annual precipitation (see p. 77). The legend is the same as that in Color Plate 22.

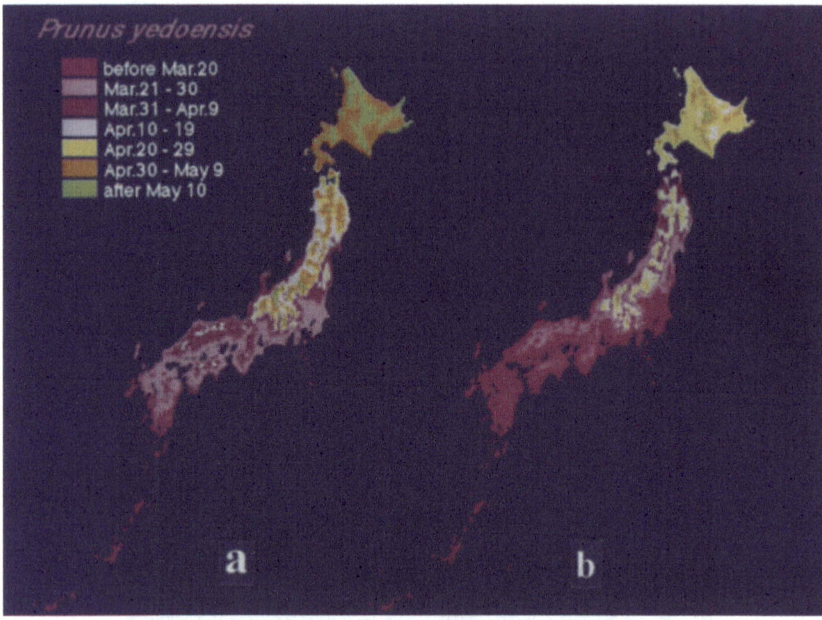

Color Plate 26 Estimation of blooming dates of *Prunus yedoensis* (see p. 87). **a**: 30-year (1951-1980) mean of blooming dates, **b**: shifted blooming dates under 3°C rise of mean temperature.

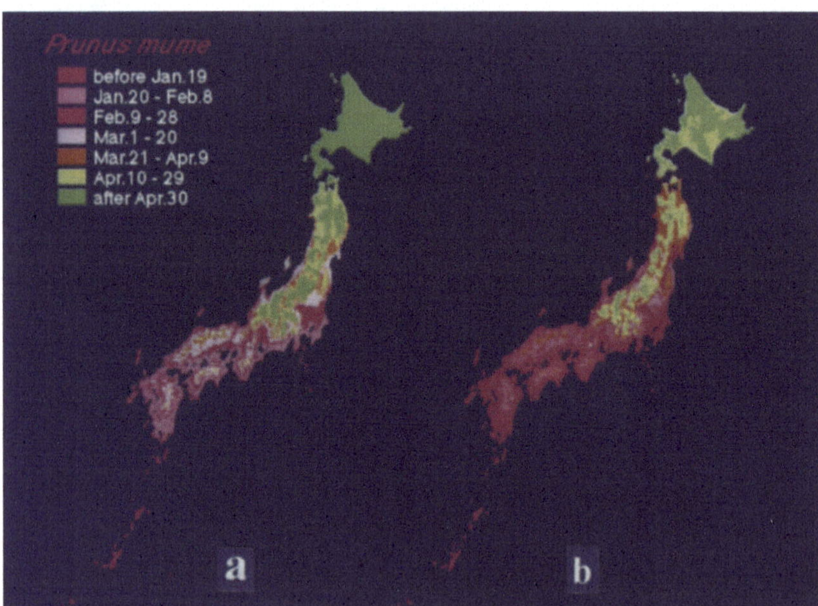

Color Plate 27 Estimation of the blooming dates of *Prunus Mume* (see p. 88). **a**: 30-year (1951-1980) mean of blooming dates, **b**: shifted blooming dates under 3°C rise of mean temperature.

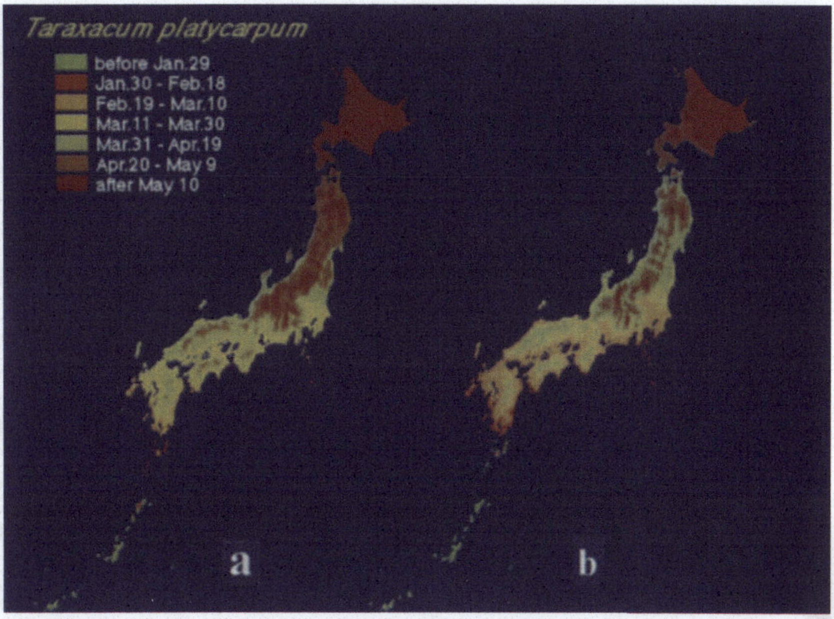

Color Plate 28 Estimation of the blooming dates of *Taraxacum* sp.(see p. 88). **a**: 30-year (1951-1980) mean of blooming dates, **b**: shifted blooming dates under 3°C rise of mean temperature.

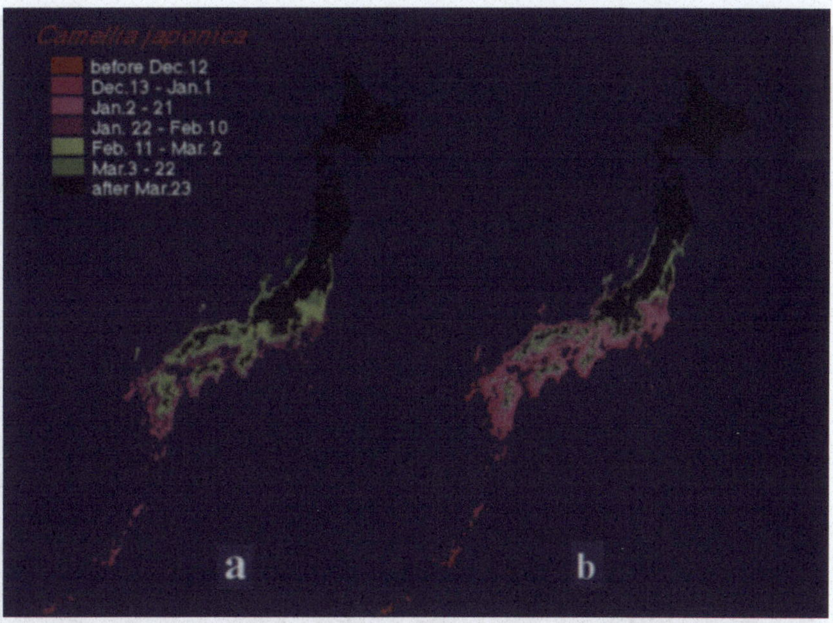

Color Plate 29 Estimation of the blooming dates of *Camellia japonica* (see p. 89). **a**: 30-year (1951-1980) mean of blooming dates, **b**: shifted blooming dates under 3°C rise of mean temperature.

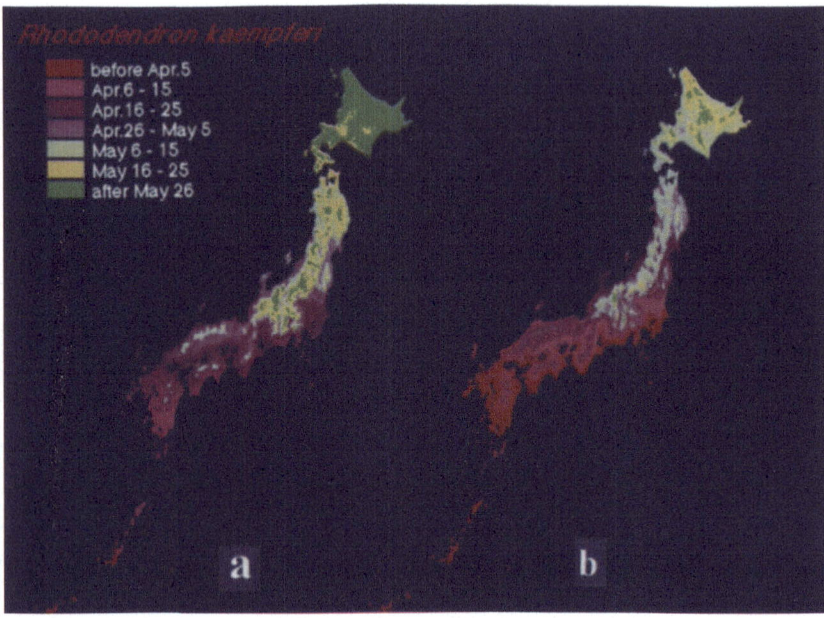

Color Plate 30 Estimation of the blooming dates of *Rhododendron Kaempferi* (see p. 89). **a**: 30-year (1951-1980) mean of blooming dates, **b**: shifted blooming dates under 3°C rise of mean temperature.

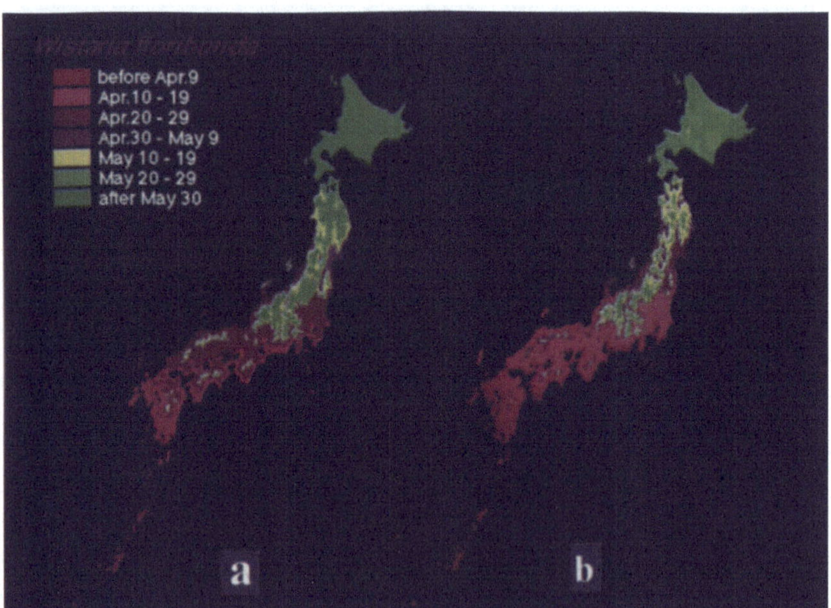

Color Plate 31 Estimation of the blooming dates of *Wistaria floribunda* (see p. 89). **a**: 30-year (1951-1980) mean of blooming dates, **b**: shifted blooming dates under 3°C rise of mean temperature.

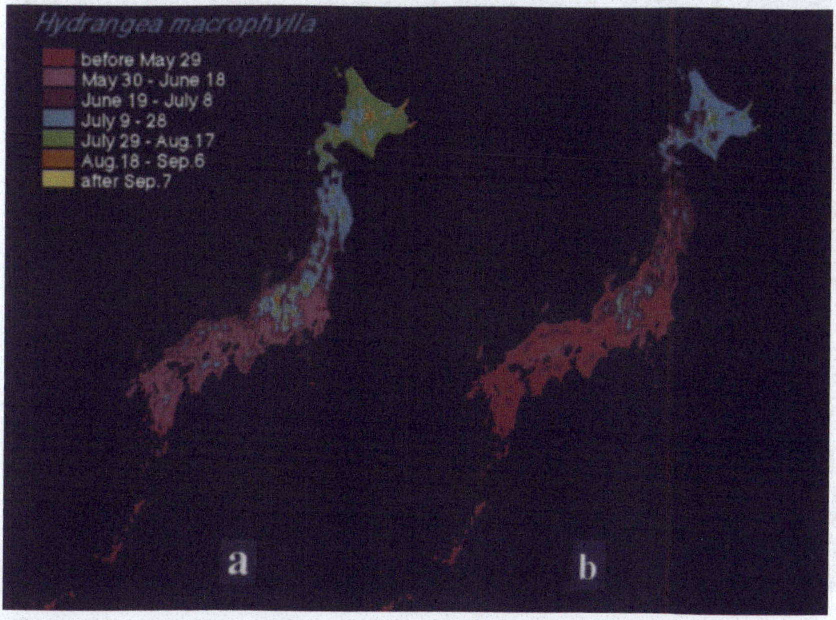

Color Plate 32 Estimation of the blooming dates of *Hydrangea macrophylla* (see p. 89).
a: 30-year (1951-1980) mean of blooming dates, **b**: shifted blooming dates under 3°C rise of mean temperature.

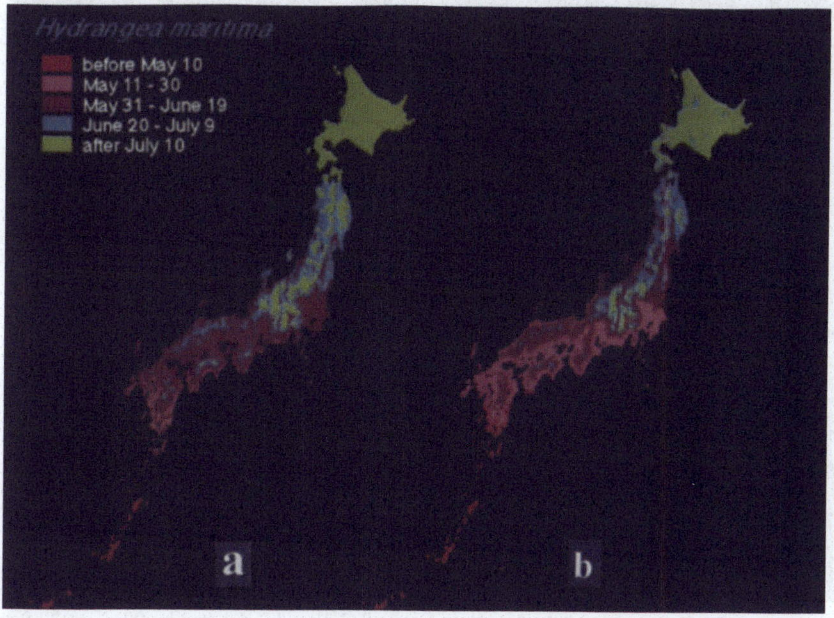

Color Plate 33 Estimation of the blooming dates of *Hydrangea maritima* (see p. 90).
a: 30-year (1951-1980) mean of blooming dates, **b**: shifted blooming dates under 3°C rise of mean temperature.

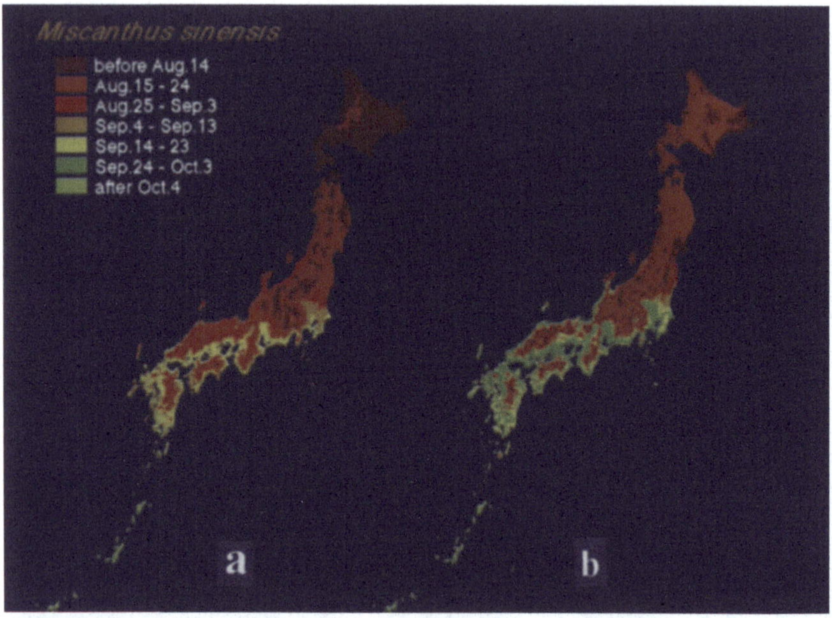

Color Plate 34 Estimation of the blooming dates of *Miscanthus sinensis* (see p. 90). **a**: 30-year (1951-1980) mean of blooming dates, **b**: shifted blooming dates under 3°C rise of mean temperature.

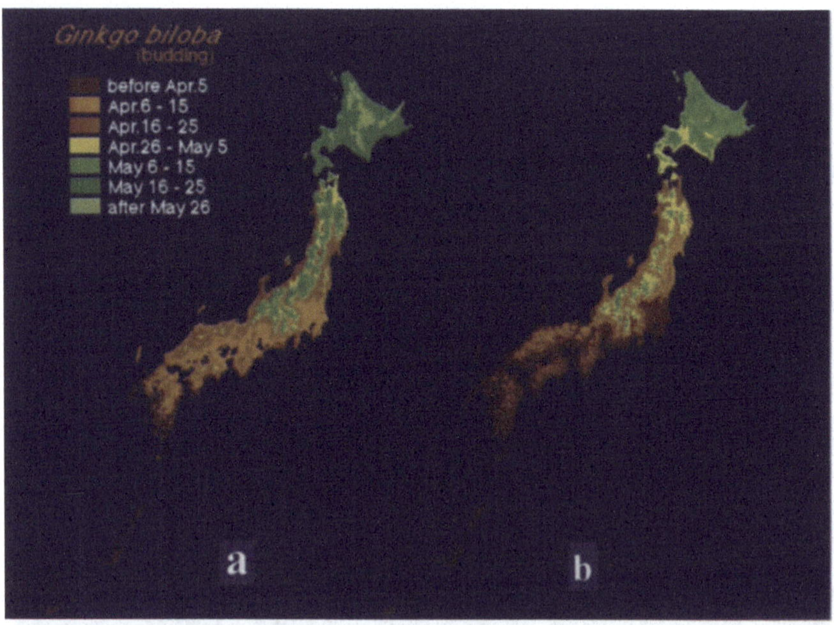

Color Plate 35 Estimation of the budding dates of *Ginkgo biloba* (see p. 90). **a**: 30-year (1951-1980) mean of blooming dates, **b**: shifted blooming dates under 3°C rise of mean temperature.

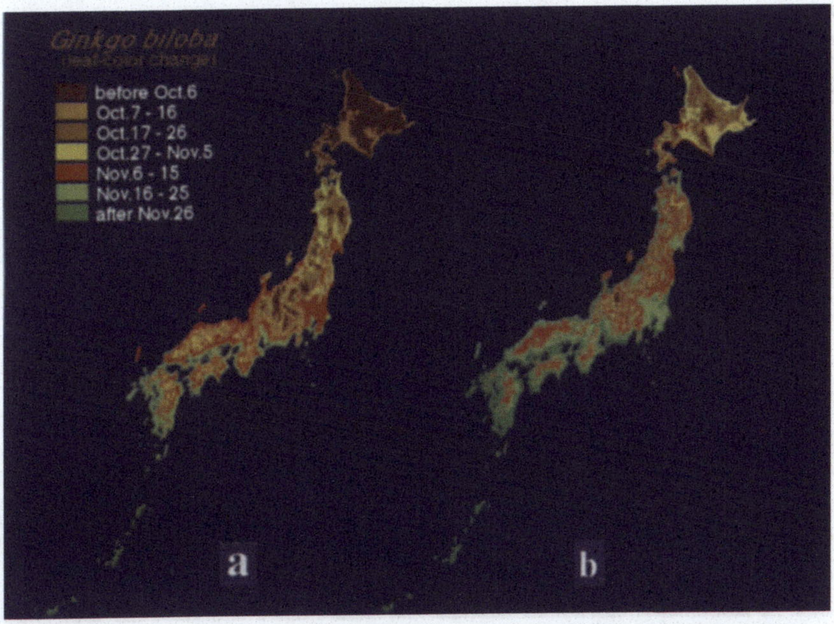

Color Plate 36 Estimation of the dates of leaf-color change of *Ginkgo biloba* (see p. 90). **a**: 30-year (1951-1980) mean of blooming dates, **b**: shifted blooming dates under 3°C rise of mean temperature.

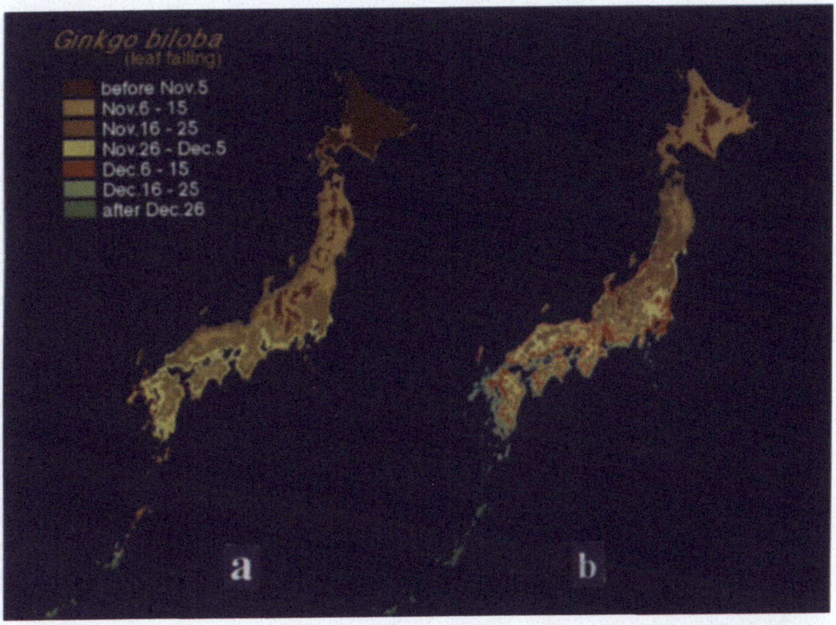

Color Plate 37 Estimation of the dates of leaf-falling of *Ginkgo biloba* (see p. 90). **a**: 30-year (1951-1980) mean of blooming dates, **b**: shifted blooming dates under 3°C rise of mean temperature.

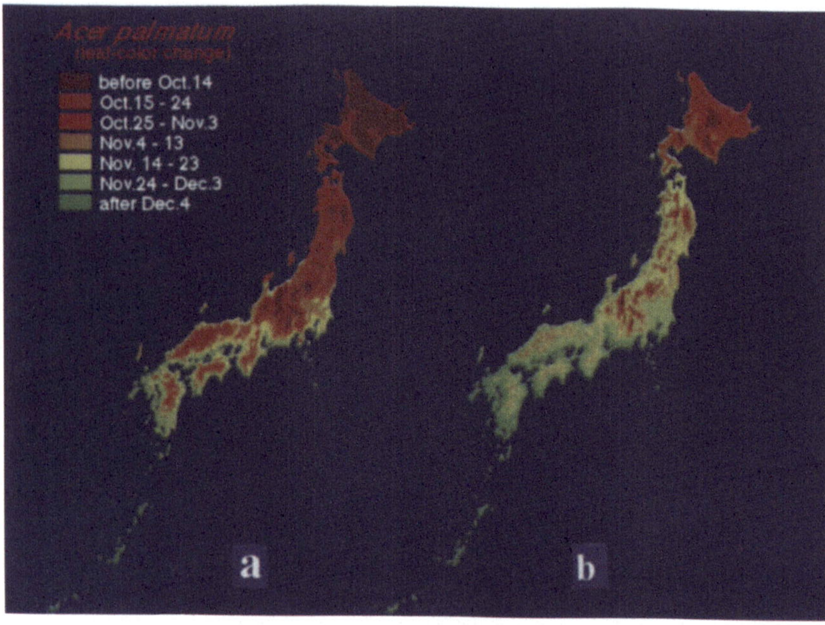

Color Plate 38 Estimation of the dates of leaf-color change of *Acer palmatum* (see p. 90). **a**: 30-year (1951-1980) mean of blooming dates, **b**: shifted blooming dates under 3°C rise of mean temperature.

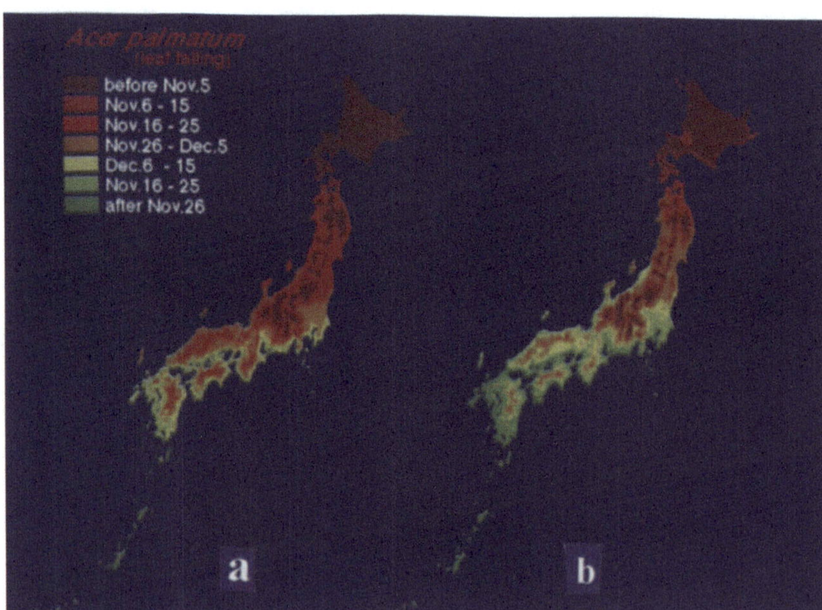

Color Plate 39 Estimation of the dates of leaf-falling of *Acer palmatum* (see p. 91). **a**: 30-year (1951-1980) mean of blooming dates, **b**: shifted blooming dates under 3°C rise of mean temperature.

Color Plate 40 The eastern slope of Mt. Zarumori viewed from an airplane, in late winter of 1994 (see p. 140). The arrow shows the study site. A snowbank forms in the leeside of Mt. Zarumori

Color Plate 41 Studied nivation hollow and meadow viewed from the north side, July 25 1992 (see p. 140). The arrow is a point on the soil profile shown in Color Plate 42.

Color Plate 42 Peat layer buried by surface of mineral-rich soil (see p. 141). To-a tephra (see text) is contained in the lowest part of the buried peat.

I . Modeling and Prediction

CLIMATIC CHANGE SCENARIOS FOR MONSOON ASIA BASED ON 2 × CO_2-GCM EXPERIMENTS

Zenbei Uchijima[1] and Shunji Ohta[2]

ABSTRACT: This investigation is made of the possible climatic change scenarios due to a doubling of carbon dioxide concentration in the atmosphere. The three global climate or general circulation models(GCM) have been used to produce the climatic change scenarios for Monsoon Asia. Using these scenarios, the information was obtained on the possible distribution of annual air temperature, annual precipitation, annual global solar radiation, annual net radiation, and annual average of radiative dryness index on the Monsoon Asia. Regional averages of relative changes in climatic elements under 2xCO_2 induced warm climates were calculated for the four sub-areas corresponding to the ice and desert climate area(I), cool and/or warm temperate climate area (II), subtropical and/or tropical climate area(III), and humid tropical area (IV). The relative changes in annual temperature were fairly large on the sub-area I than on the other sub-areas. The grid mesh maps showing the regional distribution of climatic elements under 2xCO_2 conditions can be used approximately as climatic change scenarios for Monsoon Asia.
KEY WORDS: climatic change scenario, Monsoon Asia, CO_2-doubling, GCM.

1. INTRODUCTION

All living things on the Earth are depending on photosynthates produced by green plants. As scientific researches have made clear, both vegetational and reproductive growths of plants are strongly affected by surrounding environmental conditions. Therefore, environmental conditions, particularly climate exerts the dominant control on the distribution of major vegetation types on the Earth and their productivity. This indicates that the capacity and sustainability of human use of the biosphere through agriculture, animal production and forestry are also significantly influenced by variations in weather and climate. Although the modern agriculture of developed countries is well supported by various kind of technology and much energy, it is still very sensitive or vulnerable to variations in weather conditions.

On the other hand, the uncurved growth of human population and the exponential expansion of industrial activity are causing the severe degradation of the

[1] School of Humanities, Miyazaki Municipal University, 1-1-2 Funatsuka, Miyazaki 880, Japan.
[2] School of Human Sciences, Waseda University, 2-579-15, Mikajima, Tokorozawa, Saitama 359, Japan.

global environment. Of these degradation anthropogenic change or alteration of the global climate due to the addition of greenhouse effect gases to the atmosphere would be most important for the world. Theoretical researches suggest that a doubling of carbon dioxide concentrations over preindustrial levels would cause increases in global temperatures from 1.5°C to 4.5°C, and consequent changes in rainfall, evaporation and other weather patterns. These anthropogenic climatic changes could lead to the broad-scale redistribution of terrestrial vegetation and consequent crop-belt.

Therefore the assessment of possible effects of the anthropogenic climatic changes on agricultural production and human society is needed to seek and establish the methods and policy for mitigating adverse effects of anthropogenic climatic changes. Future climate scenarios for increasing CO_2 concentration in the atmosphre provide a powerful basis for studies of impacts of climatic changes on biosphere, agriculture and society. There are two major approaches for projecting the patterns of climatic changes in a future, warmed climate (Bach, 1988):

a) analogue method based on regional and seasonal patterns of past warm climates as that in the Hypsithermal interval (BP 6000 ± 3000 yrs) and

b) physical method based on the use of scenarios from global climate models.

In the analogue method, warm climates that have actually existed are used to produce warm-world scenarios as analogues for a future CO_2-induced climate. On the other hand, the physical method uses global climate or general circulation models to produce climate scenarios with the purpose of quantitative assessment of regional and seasonal patterns of climatic changes due to the increase in greenhouse effect gases.

In this study, three physical models (GCMs), namely the Goddard Institute for Space Studies (GISS) GCM (Hansen *et al.*, 1984), the Geophysical Fluid Dynamics Laboratory (GFDL) GCM (Wetherald and Manabe, 1986), and the United Kingdom Meteorological Office (UKMO) GCM (Wilson and Mitchell, 1987) are used to produce the future climate scenarios for the Monsoon Asia under CO_2 doubling conditions.

2. CONSTRUCTION OF FUTURE CLIMATE SCENARIOS

2.1 Normal climate data used for construction of regional distribution of climate elements

The normal climate data used in this study were collected and compiled from :

a) Agroclimatic Data for Asia (FAO,1987)

b) Climate Maps of China (Chinese Central Meteorological Agency, 1978)

c) Climatic Table for the World (Japan Meteorological Agency, 1994)

d) Agroclimatic and Dry-Season Maps for South, South-East, and East Asia (International Rice Research Institute, 1982)

e) Climatic Table for Japan (Japan Meteorological Agency, 1991)

Although the time period used in calculating mean values of each meteorological element differs somewhat between each data source, it was assumed that these climatic data represent approximately the normal of each climatic element for the past 30 years. Normal climatic data so obtained (air temperature, precipitation, global solar radiation, cloud cover and vapor pressure) on Monsoon Asia (70° E to 150° E and 10° S to 50° N; land area = 1.956×10^9ha) were converted into digital form on a uniform 1° latitude by 1° longtitude meshes.

The following empirical equation (Yoshida and Shinoki, 1978) was used to estimate monthly global solar radiation (S_t ,MJ/m^2) for sites without global solar radiation measurements.

$$S_t = S_0\left[0.146 + 0.534\left(\frac{\tau}{\tau_0}\right) + 0.047\left(\frac{N}{N_0}\right) + 0.036h_\circ\right] \quad (1)$$

where S_0 is the extraterrestrial horizontal solar radiation(MJ/m^2), τ_0 and τ are the possible and actual sunshine duration (hr), respectively, N_0 and N denote, respectively, the number of whole days and days with snow cover above 10 cm of each month, and $h_\circ$ is the monthly mean of the solar altitude at the noon. The extraterrestrial horizontal solar radiation on Monsoon Asia was estimated from Iqbal's book (Iqbal, 1983). Seino (1987) has concluded from the comparison of the observed solar radiation and solar radiation estimated from Eq. (1) for 11 weather stations in Japan that Eq. (1) can be used to estimate monthly mean of daily global solar radiation with root mean square error(RMSE) of 0.333 MJ/m^2.

The following Brunt type equation was used to estimate net radiation (R_n) determining the energy level for meteorological and biological phenomena on the earth's surface.

$$R_n = (1-\rho)S_t - A\sigma T_a^4\left(0.39 - 0.058\sqrt{e_a}\right)\left(1 - cn^2\right) \quad (2)$$

where ρ is the albedo of underlying surfaces (0.0 - 1.0), A is a conversion coefficient from cal/(cm^2 min) to MJ/m^2, e_a denotes water vapor pressure (mm Hg), n is cloud amount (0.0 - 1.0), and c is a proportional constant changing from 0.5 at lower latitudes to 0.7 at higher latitudes.

The radiative dryness index (RDI) characterizing dryness or aridity of climate was calculated from

$$\mathrm{RDI} = \frac{R_n}{lr} \quad (3)$$

where R_n is the annual net radiation(J/m^2 or kcal/cm^2), l is the latent heat of vaporization of water(2.5 x 10^6 J/kg or 0.6kcal/g H_2O), and r denotes the annual precipitation (kg/m^2 or g/cm^2). As reported by Budyko (1971) and Hare (1983), there is a close correlation between radiative dryness index and vegetation type. This index is also used to calculate the net primary productivity of natural vegetation from climatic data (Uchijima and Seino, 1985 , Ohta *et al.*, 1993).

2.2 Future climate scenarios under the doubled CO_2-conditions

As already described, the three GCMs-a model developed at the Goddard Institute for Space Studies (hereafter referred to as the GISS model), one developed at the Geophysical Fluid Dynamical Laboratory(the GFDL model), and one developed at the United Kingdom Meteorological Office (the UKMO model) were selected in order to provide the future climatic scenarios for the Monsoon Asia under doubled CO_2 conditions. The physics and detailed features of these models are well described in Bach (1988) and Houghton *et al.* (1990). Table 1 shows the major characteristics of these three models.

Table 1 Summary of the three models used in equilibrium 2 x CO_2 experiments (after Houghton *et al.*, 1990)

Group	Investigators		Waves, or °lat x °long resolution	No. of vertical layers	ΔT(°C)	Δr(%)
GISS	Hansen *et al.*	1984	8° x 10°	9	4.2	11
GFDL	Wetherald & Manabe	1986	R15	9	4.0	8
UKMO	Wilson & Mitchell	1987	5° x 7.5°	11	5.2	15

ΔT is equilibrium change in global surface air temperature on doubling CO_2.
Δr is percentage change in global precipitation.

The following difference method was used to produce climatic change scenarios under 2 x CO_2 conditions:

$$M_{i,w} = M_{i,n} + \left(M_{i,2CO2} - M_{i,1CO2}\right) \quad (4)$$

where M_i denotes the *i*-th climate element, suffixes *w* and *n* show, respectively, the climatic elements under warmed and normal climates, suffixes 1 and 2 indicate GCM generated equilibrium values for 1xCO_2 (levels of pre-industrial times) and 2xCO_2, respectively. Changes in cloud amount were estimated indirectly through changes in global solar radiation projected by the GCMs, and vapor pressure under warmed climates was assumed to increase in proportion to air temperature rise.

The conversion of the normal climate data and equilibrium climate data(for 1 x CO_2 and 2 x CO_2) into the digital form on a uniform 1° latitude by 1° longitude grid meshes for the study area was made according to the following steps.

a) First, the normal values of annual mean and/or annual total of each climate element ($M_{i,n}$) for the each grid point were estimated using an interporating method and the distribution maps of normal climate elements,
b) Second, the distribution maps of ($M_{i,2CO2}$ - $M_{i,1CO2}$) were made for respective climate elements using the GCM generated data for 1 x CO_2 and 2 x CO_2 concentrations, respectively. After that, difference values ($\Delta M_i = M_{i,2CO2} - M_{i,1CO2}$) for the each grid point were estimated using an interporating method and the distribution maps of ΔM_i on the study area,
c) Lastly, $M_{i,n}$ - values and ΔM_i - values so obtained for the each grid point on the

study area were substituted into Eq. (4) to produce the grid mesh data of climatic elements for a future warmed climate.

3. PROJECTED CO_2-INDUCED CLIMATIC CHANGE SCENARIOS

3.1 Regional average of relative climatic changes

Calculations were made for the four sub-areas and the whole area (70°E to 150°E and 10° S to 50°N; 1.958 x 10^9 ha). The sub-areas I, II, III, and IV correspond to ice and desert climate areas (70°E to 100°E and 30°N to 50°N; 0.433 x 10^9 ha), cool-and/or warm-temperate climate areas (100°E to 150°E and 30°N to 50°N; 0.500 x 10^9 ha), subtropical and/or tropical climate areas (70°E to 150°E and 10°N to 30°N; 0.733 x 10^9 ha), and humid tropical climate areas (70°E to 150°E and 10°S to 10°N ; 0.292 x 10^9 ha), respectively.

Figure 1 compares climatic changes due to CO_2 doubling projected by the three GCMs for the four sub-areas and the whole of Monsoon Asia. As shown in this figure, these models generate the future climate that differ somewhat from each other. In general, the UKMO scenario produced temperature increases somewhat larger than those produced by the other two models. It was observed that the regional average of temperature increases tends to diminish gradually from the sub-area I to the sub-area IV in the all models. Namely, the regional averages of temperature

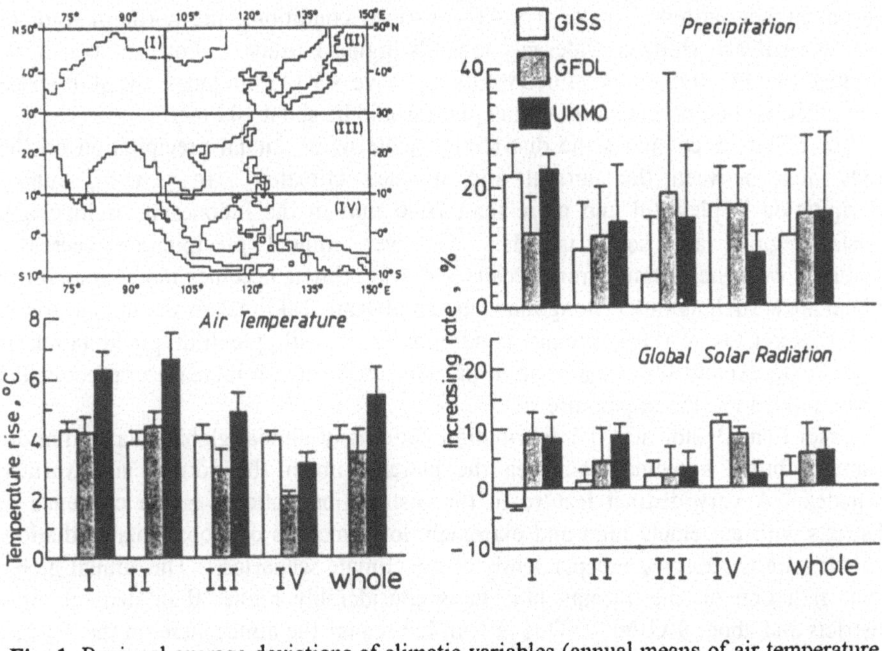

Fig. 1. Regional average deviations of climatic variables (annual means of air temperature, annual precipitation, and annual global solar radiation) under 2 x CO_2 conditions from those for the normal climate. Error bars indicates standard deviation.

increases for the sub-areas I and II were over a range from 4°C to about 6°C, while those for the sub-areas III and IV varied from 2°C to 5°C. However, the such tendency was not observed for the annual precipitation and the annual amount of global solar radiation.

Precipitation, playing an important role in plant growth and consequently biomass production, was projected to increase with CO_2-doubling. This is probably because of the increment in precipitable water in the air with temperature increases. The precipitation increment was largest in the ice and desert climate area(sub-area I) and was 10-20 % higher than that at the present. As error bars indicate clearly, the spatial variation in precipitation increment projected from the models was considerably larger compared with those for air temperature and global solar radiation. The sensitivity of global solar radiation to CO_2-doubling was not very large compared with the other two climatic elements and was between 2% and about 10 %. Although the increment in global solar radiation due to CO_2 doubling is fairly small, its increase would give important effects on biomass production and evaporation (Ohta *et al.*, 1993 ; Uchijima *et al.*, 1994).

3.2 Future climate scenarios

Future climate scenarios for Monsoon Asia under 2 x CO_2 conditions estimated by the three GCM-experiments and the methods described above are given in Color Plates 1, 2, 3, 4, and 5.

As shown in Color Plate 1, although there is broad agreement between the distribution patterns of annual mean of air temperature for higher at lower latitudes than at higher latitudes, under 2 x CO_2 warmed conditions, the northern limit of each thermal belt shifts considerably towards higher latitudes. For example, if we assumed that by 2090 temperatures would be above 3.3°C from today, the shift speed of the thermal belt on the North China plain would be about 10 km/yr.

Color Plate 2 compares the distribution patterns of annual precipitation on the study area between the normal and warmed climates. In general, annual precipitation is plentiful and more than 2000 mm in the Indonesian archipelago, coastal regions and some islands. However, annual precipitation decreases gradually with the distance from oceans and arid climate becomes more common in inland areas such as Inner Mongolia, Tibetan plateau, Taklimakan shamo and so on. Under 2 x CO_2 warmed climate conditions, area with plentiful precipitation is projected to expand somewhat than at present, reflecting an increase in precipitable water in the air with temperature rise.

Color Plate 3 indicates the distribution patterns of annual global solar radiation. There is broad agreement between the patterns under the normal and warmed climates. A very distinct feature in the distribution patterns is the existence of districts with extremely high and extremely low amounts of global solar radiation. This feature is observed independently of the climate scenarios. The annual global solar radiation on the Tibetan plateau is considerably higher than that on other districts and about 9 GJ/m^2. This is mainly because the atmosphere on the Tibetan plateau with the average altitude of about 5000 m is very thin, resulting in the higher transparency to solar radiation. On the other hand, the solar radiation in the Szuchan basin is extremely low and lower than 4 GJ/m^2. This is because of the strong interception of solar radiation by dense clouds. Namely the annual average of cloud cover on the Szuchan basin is above 8.

Color Plate 4 compares the distribution patterns of annual net radiation between the normal and warmed climates. Independently of climate scenarios, the annual net radiation is about half of the annual global solar radiation. This is due to that much energy is lost by the reflection of solar radiation and by the emission of thermal radiation from underlying surfaces into the atmosphere. The annual net radiation on areas with alpine, ice, and desert climates is negative, implying that the underlying surfaces are loosing energy by thermal radiation. With temperature rises due to doubling CO_2, the area with negative value of net radiation decreases gradually. Particularly, in the climate scenario from the UKMO-GCM, which projected temperature rise of about 6℃, the area with negative values of net radiation was smallest among the three GCMs.

Color Plate 5 shows the distribution patterns of radiative dryness index (RDI) on the study area. At first glance, there is very good agreement in the patterns between the normal and warmed climates. This implies that in this study area radiative dryness index characterizing the conversion efficiency of solar energy into biomass is not very affected by future climatic changes.

To quantify changes in the distribution patterns of each climate element on the study area, the distribution curves of percentage area with different values were calculated using the grid mesh maps shown in the color plates. Figure 2 indicates the results so obtained. As seen in this figure, the distribution curves are quite different among climatic elements. Namely, the distribution curve for annual mean of air temperature is characterized by a curve with a pointed peak over a range above 25°C and a broad skirt with the height of 5% over a temperature range below 25°C. The distribution curve of annual precipitation can be approximated by a χ^2 distribution with the degree of freedom above 2.0.

On the other hand, the distribution curves of the annual totals of global solar radiation have distinct two peaks. For the normal climate condition, the peak on a range between 6.5 and 7 GJ/m^2 reached about 15% and is two times as large as the second peak on a range between 4.5 and 5.25 GJ/m^2. The distribution curves of annual net radiation can be characterized by a curve with three peaks on a null range, on a range 1.6 to 3.0 GJ/m^2, and on a range between 4 and 5 GJ/m^2. The distribution curves of radiative dryness index have also distinct three peaks on a null range, on a range of 0.8, and on a range above 3.2. The peak on a range of 0.8 is considerably pointed.

The distribution curves under 2 x CO_2 conditions shifted towards higher value range than at present. The shift of the distribution curve is more distinct for air temperature, net radiation, and global solar radiation than for the other two elements. Although the distribution curves for radiative dryness index are not very sensitive to climatic changes, the percentage area with radiative dryness index below 1.0, in which the energy conversion efficiency would be maintained at a higher level, decreased from 56% for the normal climate, through about 53% for the GISS- and GFDL-climate scenarios, to 48% for the UKMO-climate scenario.

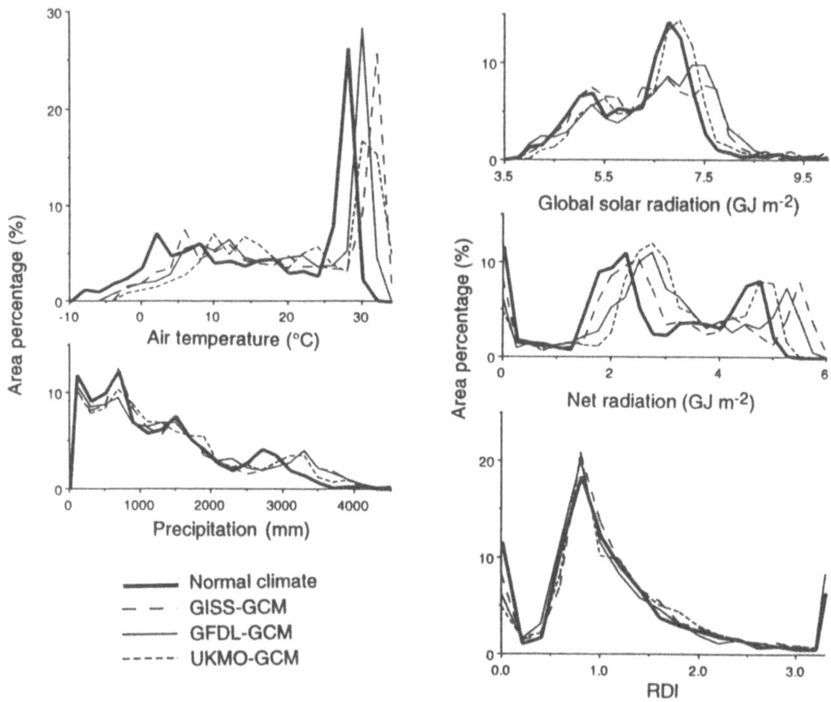

Fig. 2. Distribution curves of percentage area with different values for each climate element and for specified climatic scenarios.

4. CONCLUDING REMARKS

The normal climate data and the climate data projected by three 2 x CO_2 GCM experiments were processed to provide future climate scenarios for Monsoon Asia under 2 x CO_2 conditions. Climate values at each grid point on the study area were used to provide grid mesh maps of annual mean or annual total of five climate elements (air temperature, precipitation, global solar radiation, net radiation, and radiative dryness index). The grid mesh maps presented here can be used as a first order future climate scenario for Monsoon Asia under 2 x CO_2 conditions.

Plant growth and consequent biomass production are strongly affected by seasonal changes in climate. Particularly soil moisture affecting yield of summer crops is known to be influenced significantly by regional and seasonal changes in precipitation and evaporation. As regards precipitation, however, there is less agreement and it is pointed out that precipitation scenario obtained from GCM of the first generation is not meaningfull. Although there is such an uncertainty in precipitation scenarios, it is reasonable to assume that the grid mesh maps of climate elements presented here could be used as a first approximation of a future climate for impact assessment studies.

In order to improve our knowledge of possible regional patterns of climatic change, it is needed to study the future climate scenarios for Monsoon Asia using results of high resolution GCM experiments.

REFERENCES

Bach, W., 1988. Development of climatic scenarios-from general circulation models. In: *The Impact of Climatic Variations on Agriculture* (1). (ed by M.L. Parry, T.R. Carter and N.T. Konijn,), 125-158, Kluwer Academic Publishers.

Budyko, M.I., 1971. *Climate and Life.* (translated into Japanese by Z.Uchijima and S. Iwakiri, 1973). Tokyo University Press, 488pp.(in Japanese).

Chinese Central Meteorological Agency, 1978. *Climate Maps of China.* Geography Publishing Office, 226pp. (in Chinese).

Food and Agriculture Organization of the United Nations, 1987. *Agro-climatological Data for Asia.* FAO, 879pp.

Hansen, J., Lacis., A., Rind, D., Russel, L.,Stone, P., Fung, I., Ruedy, R. and Lerner, J., 1984. Climate sensitivity analysis of feedback mechanisms. In: *Climate Processes and Climate Sensitivity.* (ed by J. Hansen and T. Takahashi), Geophys. Monogr., Ser. 29, 130-163, Washington.

Hare, F.K., 1983. *Climate and Desertification: A Revised Analysis.* WCP-44, UNEP, 149pp.

Houghton, T., Jenkins, G.J. and Ephraums, J.J.(eds.), 1990. *Climate Change, the IPCC Scientific Assessment.* Cambridge University Press, 365pp.

International Rice Research Institute(ed.), 1982. *Agroclimatic and Dry-Season Maps of South, Southeast, and East Asia.* IRRI, 14pp. with 4 figures.

Iqbal, M., 1983. *An Introduction to Solar Radiation.* Academic Press, 390pp.

Japan Meteorological Agency, 1991. *Climatic Table for Japan* (1961-1990) (1). JMA, 478pp.(in Japanese).

Japan Meteorological Agency, 1994. *Climatic Table for the World* (1961-1990). JMA, 212pp. (in Japanese).

Ohta, S., Uchijima, Z. and Oshima, Y., 1993. Probable effects of CO_2-induced climatic changes on net primary productivity of terrestrial vegetation in East Asia. *Ecol. Res.*, **8**, 199-213.

Seino, H., 1987. Climatological calculation of solar radiation. In:*Distribution and Measurements of Solar Radiation.* (ed by K. Shibata and Z. Uchijima). 127-144, Gakkai-Shuppan Center. (in Japanese).

Uchijima, Z., Ohta, S. and Ohbuchi, Y., 1994. *Climatological Estimation of Dry Matter Production and Carbon Absorption of Natural Vegetation in the Indonesian Archipelago.* Japan International Forestry Promotion and Cooperation Center (FIFPRO). 40pp.

Uchijima, Z. and Seino, H., 1985. Agroclimatological evaluation of net primary productivity of natural vegetation (1) Chikugo model. *J. Agr. Meteorol.*, **40**, 343-352.

Wetherland, R.T. and Manabe, S., 1986. An investigation of cloud cover change in response to thermal forcing. *Clim. Change*, **8**, 5-23.

Wilson, C.A. and Mitchell, F.F.B., 1987. Simulated climate and CO_2 induced climate change over western Europe. *Clim. Change*, **10**, 11-42.

Yoshida, S. and Shinoki, S., 1978. Preparation of the maps of monthly mean global solar radiation and its year to year variability for the Japanese islands. *Tenki(Weather)*, **25**, 375-387.(in Japanese).

PROBABLE EFFECTS OF CLIMATIC CHANGES ON PLANT PRODUCTION OF MONSOON ASIA

Zenbei Uchijima[1] and Shunji Ohta[2]

ABSTRACT: This paper attempts to estimate the geographical distribution of net primary productivity (t dry matter/(ha · yr)) of natural vegetation in Monsoon Asia, in which about 45% of the world population is living at present. Estimates have also been made of probable effects of climatic changes to be induced by CO_2 increase and human land use on the total net primary production (t dry matter/yr) of this area. The Chikugo model based on relationship between climatic factors and dry matter production was used to calculate NPP values of natural vegetation. Normal climate data and climate scenarios generated by three GCMs were used to estimate to what extent plant production is affected by climatic changes due to doubled CO_2. It was expected that the potential total net production of this area will increase by about 10 % than the present level. This implies that the potential of this area for agriculture, animal production and forestry would increase consequently with the climatic changes. To make clear probable effects of human land use on the plant production, the data on land use of the Republic of Indonesia were used in cooperation with the Chikugo model. It was predicted that the actual total net production of this country is about 88% of its potential one.
KEY WORDS: NPP, global climatic change, Monsoon Asia, natural vegetation, total net production.

1. INTRODUCTION

Monsoon Asia, which lies between 10°S and 50°N in latitude and between 70°E and 150°E in longitude, is one of the most important areas in the world, because about 45% of the world population is living in this area depending mainly upon natural resources. Particularly the plant productivity of this area is surviving the each district as a source for the human use of the biosphere maintained in this area.

As well known, net primary productivity that is the net creation of organic matter by green plants using solar radiation is a key quantity for the carrying capacity of individual districts as well as the Earth. Therefore, dimension, activity and species

[1] School of Humanities, Miyazaki Municipal University, 1-1-2 Funatsuka, Miyazaki 880, Japan.
[2] School of Human Sciences, Waseda University, 2-579-15 Mikajima, Tokorozawa, Saitama 359, Japan.

diversity of ecosystems are largely affected by the net primary productivity of vegetation. Plant productivity is also a crucial variable in determining the sustainability of human use of the biosphere through agriculture, animal production, and forestry. Furthermore, the plant production of global terrestrial and aquatic vegetation is an important component affecting the biogeochemical cycle of carbon among each block in the global climate systems. Therefore, the productivity of natural and managed vegetation is a natural resources of ultimate importance for regions and Earth, ecologically and economically.

According to published vegetation maps (e.g., Walter, 1964, 1968; Fang and Yoda, 1988; Publication Committee of Chinese Vegetation, 1980), natural vegetation formations in the study area change with increasing thermal resources as defined by Warmth Index (Kira, 1945) from boreal forests in the northernmost part of Northeast China, through deciduous broad-leaved forests in the North China and evergreen broad-leaved forests in the Middle and South China, to tropical rain forests and tropical seasonal forests in the Southeast and South Asia. However, natural vegetation formations on inland districts of this area are warm temperate grassland (Nei Mongol) and semi-desert or extreme desert (Xinjian, Ungur and Gansu districts). Changes in vegetation formations in these districts are mainly because of the shortage in available soil moisture during year.

Several preliminary reports (e.g., Efimova, 1977; Uchijima and Seino, 1987; Seino and Uchijima, 1992; Ohta *et al.*, 1993) revealed that the net primary productivity (NPP) of natural vegetation of this area changes to large extent from about 1 t dry matter/(ha·yr) and less in the inland districts to about 25 t and more in the Indonesian archipelago with tropical, closed canopy forests. This geographical change in the plant productivity reflects clearly regional changes of vegetation formations and accordingly climates in the study area.

On the basis of simultaneous consideration of exchange processes of CO_2 and H_2O between vegetation and the surface air layer, we developed a semi-empirical model-Chikugo model to estimate net primary productivity of natural vegetation from climate data (Uchijima and Seino, 1985). This model was well validated by comparing the plant production data obtained by this and other methods (Uchijima and Seino, 1988a; Uchijima *et al.*, 1993; Ohta *et al.*, 1993). This model has been used successfully to study the NPP distribution on a geographical scale and to assess probable effects of climatic changes on plant production (Uchijima and Seino, 1987, 1988b; Uchijima *et al.*, 1993; Seino and Uchijima, 1992; Ohta *et al.*, 1993).

In this paper, we describe first the basic structure of the Chikugo model and estimate second the geographical distribution of NPP of natural vegetation in Monsoon Asia. Estimates are also made of probable effects of doubled CO_2 level and human land use on the total net production of natural vegetation.

2. CHIKUGO MODEL

Except little and rare organisms near hot and cold springs in deep seas, all living things on the Earth are depending upon organic compounds made by green plants. Therefore estimation or determination of the productivity of natural and managed vegetation has been one of the most important subjects in crop science, production ecology and forestry. In these studies, much attention has been concentrated on the development of methods that can estimate the net primary productivity (NPP). These methods can be divided into the following six groups (Uchijima, 1993).

a. plant ecological (or harvest) method based on the measurement of biomass at a certain time interval,

b. allometry (or relative growth) methods based on definite allometric correlation between different parts of plant,

c. canopy photosynthesis method based on the use of both light-leaf photosynthesis function and light extinction function within plant canopy,

d. climatological method based on the use of relationship between biomass production and climatic factors,

e. ecosystem method based on the use of ecosystems models connecting eco-physiological activity and environmental parameters,

f. remote sensing method based on the use of remote sensing data on vegetation obtained from satellites.

In general, the methods **a**, **b**, **c** and **d** are applied to estimate the plant productivity of vegetation on a point or local scale. On the other hand, the methods **d**, **e** and **f** are useful in estimating the plant productivity on regional and semi-continental or continental scales. With increasing interest in global environmental problems, particularly global carbon cycle, many attention has been paid to the development and application of climatological method (e.g., Efimova, 1977; Lieth, 1978) .

2.1 Basic structure of Chikugo model

By considering simultaneously exchange processes of CO_2 and H_2O between plant canopy and the surface air layer, and heat balance equation of vegetation land, we obtained the following relation (Uchijima and Seino, 1985)

$$\mathrm{NPP} = \frac{A_0 R_n}{d\,(1+\beta)} \tag{1}$$

where NPP is net primary productivity (t dry matter/(ha · yr)), R_n is annual net radiation (kcal/cm^2), d is annual mean of water vapor deficit (mm Hg), β is annual mean of Bowen ratio and A_0 is numerical constant relating with diffusion resistance and CO_2 concentration in and above plant canopy.

Equation (1) can be simplified by adopting energy conversion efficiency ($\alpha = A_0 / [\,d\,(1+\beta)]$) as follows:

$$\mathrm{NPP} = \alpha\, R_n \tag{1a}$$

The above relation shows that when α value is maintained at a certain level, NPP would increase proportionally to annual net radiation, and that the magnitude of α decreases rapidly with increasing d and β characterizing dryness of climate. That is, it is expected that the energy conversion efficiency (α) decreases drastically as climate becomes dryer.

As already described in the preceding paper (Uchijima and Seino, 1985), the dependence of α on the dryness of climate was well approximated by

$$\alpha = 0.29 \exp(-0.216 \, \mathrm{RDI}^2) \tag{2}$$

where RDI ($=R_n/lr$) is radiative dryness index, l is latent heat of water evaporation (kcal/g H_2O) and r is annual precipitation (cm). Substituting the above relation into Eq.(1a) yielded

$$\mathrm{NPP} = [0.29 \exp(-0.216 \, \mathrm{RDI}^2)] \, R_n \tag{3}$$

This is the Chikugo model for estimating net primary productivity (NPP) of natural vegetation from climate data. As can be seen in Eq.(3), this model considers simultaneously effects of both dryness of climate and radiation on plant production.

2.2 Validation

To verify the applicability of the Chikugo model (Eq.3), the total net production (TNP, t dry matter/yr) of the individual prefectures of Japan was calculated from the following relation using the NPP-distribution map made by the Chikugo model and the normal climate data of Japan.

$$\mathrm{TNP}_i = \sum_{j=1}^{5} A_{ij} \, E_{ij} \, \overline{\mathrm{NPP}}_i \tag{4}$$

where TNP_i is TNP of i-th prefecture, $\overline{\mathrm{NPP}}_i$ is mean NPP of the i-th prefecture determined graphically on the NPP-map, E_{ij} and A_{ij} is production efficiency and land area for the j-th land use class in the i-th prefecture, respectively. Referring to Iwaki's report (Iwaki, 1981) the following values were used as the production efficiency for the each land use class.

$j =$ 1 (forests)	$E = 1.00$
$j =$ 2 (permanent crop fields)	$E = 0.80$
$j =$ 3 (usual crop fields)	$E = 0.81$
$j =$ 4 (grass land)	$E = 0.60$
$j =$ 5 (other land use)	$E = 0.00$

Using the data on the normal climate and land use of the each prefecture, and the above production efficiency, TNP values were estimated from Eq. (4). On the other hand, Iwaki (1981) estimated the TNP values of the each prefecture using the data of the National Census of Green Resources of Japan conducted by the Japan Environmental Agency. The TNP values estimated by these two different methods were compared with each other (Fig. 1a). The NPP map on Monsoon Asia constructed using the NPP values estimated by the Chikugo model was superimposed onto the vegetation formation maps proposed by several authors (e.g., Walter, 1964, 1968, Fang and Yoda, 1988) to estimate the average NPPs of major vegetation formations in this area. The average NPPs so obtained were compared with those proposed by Whittaker and Likens (1975) (Fig. 1b). As shown in Figs. 1a and b, there is good agreement between the production data obtained by quite different methods. This agreement indicates evidently that the Chikugo model can be applied to estimate NPP and/or TNP of natural vegetation on regional and continental scales from climatic data with acceptable error.

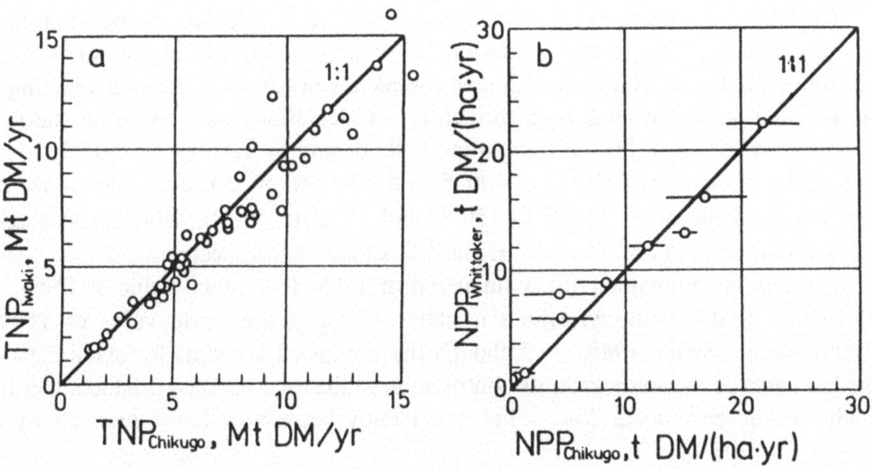

Fig. 1. a: Comparison of total net production of each prefecture under the normal climate estimated from the Chikugo model and Eq. 4 and those obtained by Iwaki (1981) using the data of National Census of Green Resources of Japan. **b:** Comparison of mean NPP of major vegetation formations under the normal climate estimated from the Chikugo model and those obtained by Whittaker and Likens (1975) using IBP-data. Horizontal bars show standard deviation.

3. RESULTS AND DISCUSSION

3.1 Net primary productivity and total net production under normal climate conditions

Calculations were made with grid squares of 1.0° latitude by 1.0° longitude. The gridded normal climate data of the study area were converted on a computer to NPP values under the normal climate conditions. The NPP data so obtained were used to provide a NPP map for Monsoon Asia. Color Plate 6 shows the geographical distribution of NPP of natural vegetation in the study area. As can be seen in this figure, the southern and eastern coastal districts of the study area and Indonesian archipelago are more productive than the inland districts with dryer and colder climate. Higher NPP values in these productive districts are mainly due to favorable combination of climate factors, that is, plentiful rainfall and higher temperature throughout year. Particularly, NPP of Indonesian archipelago, which is clothed in the dense canopy of tropical rain forests, is above 20-25 t dry matter /(ha · yr). This archipelago is one of the world most productive areas such as Amazonian basin and tropical central Africa. On the other hand, the inland district and higher mountain and plateau districts with dry and cold climate are less productive compared with forest zones of the coastal districts and tropical Indonesian archipelago. This is mainly because of the unfavorable combination of climatic factors in these districts. That is, in summer season hot and dry weather prevails in these districts and in winter season cold and dry weather prevails.

The NPP map shown in Color Plate 6 was utilized to calculate the potential total net production (TNP_0, t dry matter/yr) of natural vegetation of the study area, assuming that the study area was clothed completely in natural vegetation adapting to climate and soil conditions of each district. Calculations were made on the four sub-areas as shown in Fig. 2. Sub-areas I, II, III and IV correspond roughly to ice and desert climate area (70° E to 100° E and 30° N to 50° N), cool- and/or warm-temperate climate area (101° E to 150° E and 30° N to 50° N), sub-tropical and/or tropical climate area (70° E to 150° E and 10° N to 29° N), respectively.

The results so obtained are summarized in Table 1, the total value of $TNP_{0.n}$ is 23.26×10^9 t dry matter/yr which is about 17 % of the world value of $TNP_{0.n}$ (Uchijima and Seino, 1988a). Although the sub-area I accounts for about 22% of the land area of the study area, its contribution to the total biomass production of the study area is only about 2%. This is evidently because of lower productivity of

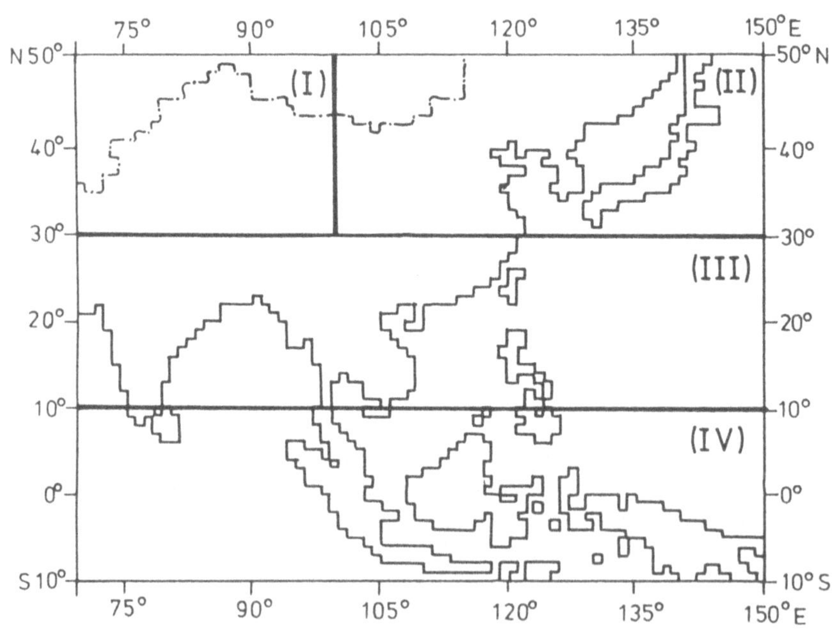

Fig. 2. Geographical distribution of study sub-areas I, II, III, and IV.

Table 1 Potential net production (TNP_0) under normal and warm climates

Sub-area	A, 10^9 ha	$TNP_{0.n}$ 10^9 t dry matter /yr	$TNP_{0.w}$, 10^9 t dry matter /yr GISS	GFDL	UKMO
I	0.433	0.474	0.662	0.689	0.721
II	0.500	3.75	4.360	4.140	4.510
III	0.733	11.22	11.900	12.600	12.400
IV	0.292	7.76	9.380	9.060	8.370
Total	1.958	23.26	26.300	26.500	26.100

plants grown on the inland district with dryer climate. On the other hand, the sub-areas III and IV with subtropical and/or tropical forests account for about 82 % of the total biomass production of the study area.

3.2 Probable effects of climate warming induced by doubled CO_2 level on NPP and TNP

Plant growth is very sensitive to variations in climate. Therefore, it is expected that climate change projected due to increased CO_2 in the atmosphere would have a significant influence on terrestrial vegetation, causing large changes in net primary productivity and vegetation distribution pattern. The three gridded warmed climatic data sets based on GISS-, GFDL-, and UKMO-global climate models were used to estimate effects of climatic warming due to doubled CO_2 level on NPP and TNP. Color Plate 7 compared the NPP-distribution patterns under the respective warmed climate scenarios.

Comparing the NPP-maps presented in Color Plate 7 and one in Color Plate 6 yields the following conclusions:
(1) the areas with higher NPP values shift northward from the present position responding to projected climatic changes, and
(2) the latitudinal and altitudinal shift of iso-NPP-zones are more pronounced in the present less productive areas than in the present more productive areas.

The distribution maps of NPP over Monsoon Asia shown in Color Plates 6 and 7 were used to assess the change in area of zones with various NPP values due to variations in climate. All calculations were made with the sub-areas I, II, III, IV and the whole area. The thin line in Fig.3 shows the distribution curve of percentage area for the normal climate. The thick line in Fig.3 denotes the distribution curve of percentage area for specified climate scenarios. As shown in Fig. 3, the frequency curves of NPP values in this study area can be classified into the three groups such as the sub-area I type with a very high peak in the null-NPP range, the sub-area II and III type with a relatively low peak and broad skirt in the NPP-values and the sub-area IV type with a pointed peak and narrow skirt in the NPP range 20 to 30 t/(ha · yr). The difference of the frequency curves between the each sub-area resulted from the difference of vegetation formations and consequently climatic conditions between them. That is, the very high peak in the null-NPP

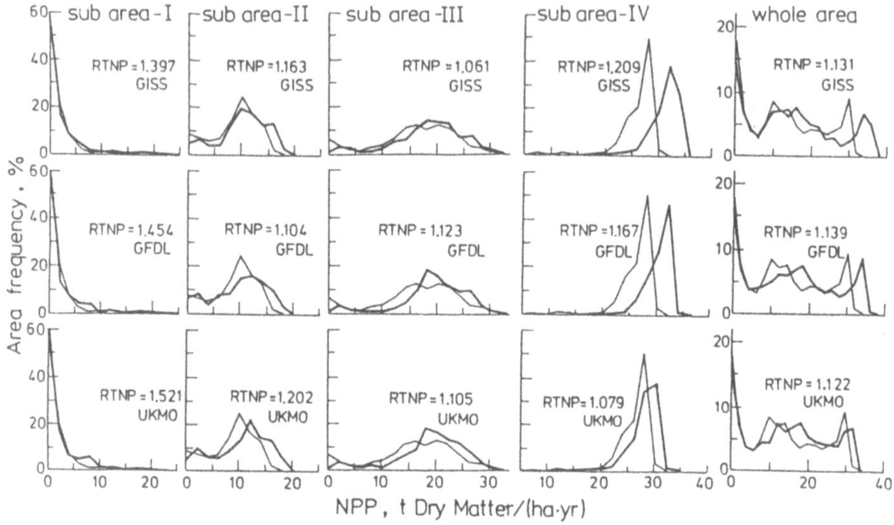

Fig. 3. Distirbution curves of percentage area with different NPP values for specified climatic scenarios. Thin and thick lines represent distributions for the normal climate and for the warmed climates, respectively.

range of the sub-area I type reflects the existence of extremely dry and cold districts in the sub-area I. The pointed peak in the NPP range 25 to 30 for the sub-area IV type is mainly due to tropical rain forests grown on the Indonesian archipelago. The NPP frequency curve for the whole area can be characterized by a curve with distinct three peaks in the null range, 10 to 20 range, and 30 range of NPP-values, reflecting the individual feature of the NPP frequency curves for the each sub-area. As can be seen in this figure, the thick line curve shifts generally towards a higher NPP range with concomitant decrease in the area with lower values of NPP.

To quantify the effect of climatic changes on the total net production (TNP_0) of the each sub-area and the whole area, the relative total net production (RTNP) was calculated from:

$$RTNP = \frac{TNP_{0.w}}{TNP_{0.n}} \tag{5}$$

where $TNP_{0.n}$ is the total net production under the normal climate conditions, and $TNP_{0.w}$ is the total net production for specified climatic scenarios. As indicated in Fig. 3, the values of RTNP range from 1.060 for the sub-area III under the GISS scenario to 1.522 for the sub-area I under the UKMO scenario. However the change in the RTNP values for the whole area is not very large and those are between 1.125 for the UKMO scenario and 1.142 for the GFDL scenario. If natural vegetation could adapt to so rapid shift of climatic zones through their fast migration, it is expected that the potential net production of natural vegetation in the study area would increase by about 10% compared with that under the normal climate conditions.

However if we assumed that CO_2 doubling will have occurred by the year 2050, the shift speed of climatically suitable habitats for tree species would be about 10 km/yr. This estimated shift speed of major tree habitats is very large compared with the migration speed of major tree species obtained by the analysis of fossil pollen for the post glacial period (Huntley and Birks, 1983). Therefore, the adaptation of terrestrial vegetation to estimated shift of climatically suitable habitats for tree species could be greatly delayed, resulting in the decline of some plants and a consequent reduction in biomass production.

3.3 Probable effects of human land use on TNP of the Republic of Indonesia

Although at present agricultural land constitutes only a small fraction (8.7%) of the global land area (ca. 14.9×10^9 ha), it was originally very fertile lands, which were clothed in closed canopy forests and well grown grass. These fertile lands have been converted into arable lands to produce foods and feeds. The conversion from forests and grass lands into arable lands induced the drastic changes in biomass production of terrestrial vegetation. Urbanization, industrial development, and construction of transportation networks have also required much space. It is known that the land area per capita required for those purposes is about 0.1-0.25 ha (Norse *et al.*, 1992). Extensive land area used for those purposes has mainly come from forest, grass lands, and agricultural lands. Therefore, human land use for agriculture, urban/industrial development, and transportation construction has brought about the significant decrease in the area of forests and consequently biomass production of terrestrial vegetation.

As already pointed, the forests of Southeast Asia are converted rapidly into arable lands, urban area and industrial area, because of the rapid population growth and quick urbanization in this area. Therefore closed tropical rain forests with higher productivity are deforested with a surprising speed.

Equation (4) was used to estimate effects of these massive deforestation on the TNP of the Repulic of Indonesia. In the calculation, the land area of the Republic of Indonesia was divided into the two groups with E_1=1.00 (forests) and E_2=0.70 (non forests). The results so obtained for major islands of the Republic of Indonesia are summarized in Table 2 (Uchijima *et al.*, 1994).

As seen from Table 2, the actual total net production (TNP_a) calculated from Eq.(4) and the data on land use is lower than the potential net production (TNP_0). Relative total net production ($RTNP=TNP_a/TNP_0$) varies from 0.722 of densely populated Java island, through 0.833 of Vali island, to 0.945 of sparsely populated Irian Jaya, with the national average of 0.886. This national average of 0.886 is fairly high compared with the preliminary estimates for Japan (0.75) and for the whole continents (0.60). The above result indicates that the effect of human land use on the plant production of the Republic of Indonesia is still not very significant at present. However it is expected that the relative total net production of this country would decrease more rapidly in the near future, leading to considerable reduction in

Table 2 Potential net produciton (TNP_0) and actual net produciton (TNP_a) of major islands of th Republic of Indonesia

	percent area (%)		TNP_0	TNP_a	RTNP
	Forests	Non-forests	10^9	t dry matter/yr	
Sumatra	43	57	1.210	1.000	0.833
Kalimantan	63	37	1.590	1.410	0.887
Sulawesi	52	48	0.456	0.405	0.888
Maluku islands	70	30	0.070	0.064	0.914
Irian Jaya	82	18	1.270	1.200	0.945
Bali	23	77	0.012	0.010	0.833
Java	7	93	0.036	0.026	0.722
Other small islands	60	40	0.155	0.137	0.884
Total or Average			4.798	4.251	0.886

the biomass and total net production of this country. Similar results are also expected for other countries in Southeast Asia.

4. CONCLUDING REMARKS

The aim of this paper has been to calculate the biomass production of natural vegetation in the Monsoon Asia and to estimate probable effects of CO_2 induced climate changes on the biomass production. Calculations were also made of effects of human land use on the total net production of the Republic of Indonesia. The results obtained in the study can be summarized as follows:

(1) The potential total net production (TNP_0) of this study area (10° S to 50° N in latitude and 70° E to 150° E in longitude) was calculated to be 23.26×10^9 t dry matter/yr by the Chikugo model, assuming that whole land area is completely clothed in natural vegetation such as grass and forests. Sub-areas III and IV, on which subtropical and tropical rain forests are well grown, were expected to account for about 80% of the whole total net production.

(2) Although the sub-area I accounts for about 22% of the whole land area of this study area, its contribution to the whole total net production is remarkably lower and about 2%. This is evidently because the natural conditions in this area are quite unfavorable to plant growth due to cold and dry climate.

(3) Climate scenarios generated by three GCMs (GISS, GFDL, and UKMO) for doubled CO_2-level were used to estimate probable effects of climate changes on the total net production. If natural vegetation could adapt to rapid shift of climatically suitable habitats for each species through the fast migration of plant species, it was expected that the total net production of this study area would increase by about 10 % due to climatic changes under doubled CO_2-level.

(4) To estimate probable effects of environmental degradation due to human land use on the total net production, calculations were made of the Republic of Indonesia using

the land use data of this country. It was found that at present the total net production of this country accounts for about 88% of its potential total net production. Although this percentage value is higher than those values for Japan (75%) and for the world (60%), it is expected that this higher percentage would decrease rapidly in the near future because of considerable population growth and so rapid economical development of this country.

REFERENCES

Efimova, N.A., 1977. *Solar Energy and Plant Productivity* (translated into Japanese by Z. Uchijima, 1994). Association for International Cooperation of Agr. and Forestry. No.14, 132pp. (in Japanese).

Fang, J.Y. and Yoda, K., 1988. Climate and vegetation in China(1). *Ecol. Res.*, **3**, 37-51.

Huntley, B. and Birks, H.J.B., 1983. *An Atlas of Past and Present Pollen Maps for Europe: 0-1300 years ago.* Cambridge Univ. Press, 173pp.

Iwaki, H., 1981. Geographical distribution of phytomass on Japan. *Environ. Inf. Sci.*, **40**, 54-60. (in Japanese).

Kira, T., 1945. *New Method for Classifying Climate of East Asia.* Kyoto Univ., Laboratory of Horticulture, 23pp. (in Japanese).

Lieth, H., 1978. Vegetation and CO_2 changes. In: *Environ. Carbon Dioxide, Climate and Society* (ed. by Williams), 103-109. Pergamon Press.

Norse, B., James, C., Skinner, B.J. and Zhao, Q., 1992. Agriculture, land use and degradation. In: *An Agenda of Science for Environment and Development into 21st Century* (ed by J.C.I. Dooge, G.T. Goodman, J.W.M. la Rivere, J. Marton-Lefevre, T. O'Riordan and F. Praderie), Cambridge Univ. Press, 331pp.

Ohta, S., Uchijima, Z. and Oshima, Y., 1993. Probable effects of CO_2 induced climatic change on net primary productivity of natural vegetaton in East Asia. *Ecol. Res.*, **8**, 199-213.

Publication Committee of Vegetation Map of China, 1980. *Vegetation Map of China.* Sci. Publishing Office, 1375pp. (in Chinese).

Seino, H. and Uchijima, Z., 1992. Global distribution of net primary productivity of terrestrial vegetation. *J. Agr. Meteorol.*, **48**, 38-48.

Uchijima, Z., 1993. Climate and vegetation. *J. Geogr.*, **102**, 745-762. (in Japanese).

Uchijima, Z., Ohta, S. and Ohbuchi, Y., 1994. Climatological estimation of dry matter production and carbon absorption of natural vegetation in the Indonesian archipelago. *JIFPRO*, 40pp.

Uchijima, Z., Ohta, S., Seino, H., Oshima, Y. and Urabe, T., 1993. Climatological assessment of potential carbon absorption by natural vegetation in monsoon East Asia. *JIFPRO*, 60pp.

Uchijima, Z. and Seino, H., 1985. Agroclimatic evaluation of net primary productivity of natural vegetation (1) Chikugo model. *J. Agr. Meteorol.*, **40**, 343-352.

Uchijima, Z. and Seino, H., 1987. *Maps of Net Primary Productivity of Natural Vegetation on Continents.* Special Report of NIAES and Kyushu Agr. Exp. Stat., 121pp.(in Japanese with English summary).
Uchijima, Z. and Seino, H., 1988a. An agroclimatic method of estimating net primary productivity of natural vegetation. *JARQ,* **21**, 244-249.
Uchijima, Z. and Seino, H., 1988b. Probable effects of CO_2-induced climatic change on agroclimatic resources and net primary productivity in Japan. *Bull. Natl. Inst. Agro-Environ. Sci.,* No.4, 67-83.
Walter, H., 1964. *Die Vegetation der Erde 1.* Gustav Fischer, 592pp.
Walter, H., 1968. *Die Vegetation der Erde 2.* Gustav Fischer, 1001pp.
Whittaker, R.H. and Likens, G.E., 1975. The biosphere and man. In: *Productivity of the Biosphere* (ed by H. Lieth and R.H. Whittaker), 305-328, Springer-Verlag.

MODEL EXPECTATION OF IMPACTS OF GLOBAL CLIMATE CHANGE ON BIOMES OF THE TIBETAN PLATEAU

Xinshi Zhang[1], Dianan Yang[1], Guangsheng Zhou[1], Chunying Liu[1] and Jie Zhang[1]

ABSTRACT: Tibetan Plateau with the average altitude of about 5000m is one of very important regions in the world, because its natural conditions are very severe and much vulnerable to changes in climate. Therefore it is expected that changes in climate and consequent biomes on this region may play a role of pilot and/or pioneer in monitoring the global environmental degradation due to intensive human activity. In this paper, the responses of biomes on the Tibetan Plateau to the global climate change induced by doubled CO_2 were studied using improved Holdridge life zone classification system, vertical vegetation belt system, permafrost model, and NPP model. The global climate changes will cause the considerable changes in the vertical distribution of vegetation, permafrost zone, and NPP-values on the Tibetan Plateau. It was expected that the increment in annual mean temperature by 4℃ and annual precipitation by 10% under doubled CO_2 condition would accerelate the speed of desertification which is spreading on this area mainly due to the intensive mass land use at present.

KEY WORDS: global climate change, Tibetan Plateau, Holdridge life zone classification system, vertical vegetation belt, NPP-model, permafrost model

1. INTRODUCTION

The exponential growth of global population and the great expansion of human activity are causing significant changes in natural conditions of the Earth. In these significant changes, the increase of greenhouse effect gases in the atmosphere is expected to have very important implication in the broad scale distribution of natural vegetation and agriculture. The Tibetan Plateau with the average altitude of about 5000m is an area which is very sensitive and vulnerable to global climate change, that is climate warming. The vegetation distribution in Tibetan Plateau is expected to change vertically with the climate warming, resulting in the altitudinal shift of vegetation zone. This is mainly because Tibetan Plateau is located in marginal land areas in which the growth and distribution of plants depend closely on local climate conditions. This implies that the natural and managed vegetation on Tibetan

---

[1] Institute of Botany, Chinese Academy of Sciences, 141 Xizhimenwai Avenue, Beijing, P.R. China.

Plateau could use as an indicator or pilot for monitoring the global climate change due to the explosing human activity and its influence on terrestrial ecosystems. In this report, several biometeorological methods such as the Holdridge life zone classification system, montane vegetation belt system, permafrost model, and improved NPP-model were used to study the response of natural vegetation in the Tibetan Plateau to global climate change.

2. CLIMATE-VEGETATION INTERACTION PATTERN ON THE TIBETAN PLATEAU

2.1 Vegetation type and climatic condition

The vegetation zonation on the Tibetan Plateau can be classified into the 11 subzones as shown in Table 1 (Zhang and Liu, 1994). The climatic conditions characterizing the individual vegetation subzones are summarized in Table 1. As can be seen in Table 1, the vegetation subzones on the Tibetan Plateau change along

Table 1 Vegetation subzones on the Tibetan Plateau in relation to climatic conditions

Vegetation subzones	BT(℃) Mean σ	PET(mm) Mean σ	P(mm) Mean σ	PER Mean σ
1	12.9±3.9	764.9±230.9	1328.0±519.0	0.6±0.2
2	8.3±3.1	488.4±182.4	701.1±240.3	0.8±0.5
3	3.5±1.0	207.2± 58.5	572.9±106.7	0.4±0.1
4	5.4±1.9	319.5±113.7	372.1±108.9	0.9±0.4
5	5.6±1.9	331.4±111.3	427.0±119.2	0.8±0.4
6	1.9±0.5	114.1± 26.4	336.3± 78.1	0.3±0.1
7	2.2±1.2	127.1± 69.6	285.0±138.1	0.6±0.7
8	3.4±2.2	199.8±128.7	102.2± 80.2	2.1±1.6
9	0.7±0.6	42.0± 30.9	92.3±130.6	0.9±0.9
10	4.6±2.3	269.7±137.5	52.0± 23.1	5.1±0.4
11	6.3±1.0	370.4± 57.4	87.8± 69.6	8.6±6.9

Note: BT: biotemperature, PET: potential evapotranspiration,
P: precipitation, PER: potential evapotranspiration rate,
σ : standard deviation,

Vegetation subzone codes are as follows:

1. Tropical Mountain Forest Subzone of the Eastern Himalaya,
2. Subtropical Mountain Forest Subzone of Western Sichuan Province and Southeastern Xizang Province,
3. Alpine Meadow and Scrubland of Eastern Xizang,
4. Temperate Steppe and Shrubland of Southern Xizang,
5. Temperate Steppe of Eastern Qinghai-Qilian Mountains,
6. Southern Qinghai Plateau Alpine Steppe,
7. Qiangtang Plateau Alpine Steppe,
8. Ngari Mountain Temperate Desert,
9. Karakoram-Northern Qiangtang Alpine,
10. Kunlun Desert Mountains,
11. Chaidam Temperate Desert.

thermal gradient from south to north and along moist gradient decreasing from southeast to northwest.

Through many climatological researches, the following regression equations were obtained between climatic parameters and geographical parameters:

$$\left.\begin{array}{ll} \mathrm{BT} = 46.18 - 0.44965L - 0.13627G - 0.0037006H & R = 0.942 \\ \mathrm{PET} = 2727.5 - 26.486L - 8.086G - 0.218432H & R = 0.942 \\ \mathrm{PER} = 10**(4.7241 + 0.041754L - 0.05083G - 0.00040665H & R = 0.757 \end{array}\right\} \quad (1)$$

where L, G, and H denote, respectively, the latitude, longitude, and altitude of each study site. The above regression equations indicate that BT and PET would change about 0.45 ℃ and 26.5mm per unit increment or decrease in the latitude, respectively. On the other hand, the change in BT and PET per unit increment or decrease in the longitude is less than for the latitude. They are, respectively, about 0.14℃ and 8.1mm. The changes of BT and PET per 100m increment or decrease in the altitude were, respectively, about 0.37℃ and 21.8mm.

2.2 Montane vertical vegetational belt system

As well known, the vegetation belt on the Tibetan Plateau changes evidently with the altitude, reflecting the vertical change in climatic conditions. As expected from Eq.(1), thermal resources on this plateau decrease with the increment in the latitude from south to north. Because the amount of precipitable water in the air decreases gradually with the increment in the distance from oceans, the precipitation on Tibetan Plateau decreases from south to north. Reflecting the vertical and horizontal changes of climatic conditions on this plateau, the natural vegetation belts of Tibetan Plateau change from forests in the south lower region, through shrub-steppe, and alpine steppe in the middle region, to alpine desert and temperate desert in the north higher regions.

The species complexity of each vegetation belt on the Tibetan Plateau is known to be more simple for the alpine desert and temperate desert in the north higher regions than for the forest zone in the south lower regions. The change of vegetation belts along the longitudinal section of 87°E is shown Color Plate 8a and 8b.

South Himalayas can be classified into subtropical marine montane type, in which 9 vegetation belts: montane rain forest and seasonal rain forest (0-1,000m), montane evergreen broadleaved forest (1,000-2,000m), montane semievergreen broadleaved forest (2,000-2,500m), montane mixed coniferous-broadleaved forest (2,500-3,000m), montane coniferous forest (3,000-3,700m), montane shrub meadow (3,700-4,500m), alpine desert (4,500-5,200m), subnival belt (5,200-6,200m), and nival belt (5,500-8,848m): distribute vertically from the lowest region to the top region.

Gangdise mountains belong to the continental semiarid montane type. Four vegetation belts such as montane shrub-grass tussock or shrub-steppe (4,100-4,450m), alpine steppe (4,450-4,950m), alpine meadow (4,950-5,150m), and subnival zone (5,150-5,950m) are distributed vertically on the south slope of this mountains. There are three vegetation belts, which are typical montane steppe (4,500-4,900m), montane desert steppe (4,900-5,300m), and subnival belts (5,300-5,950m) on the north slope.

Hohxil mountains are classified into continental semiarid montane type. The vertical zonalization of vegetation type on the south and north slopes of this mountains is quite similar to that for Gangdise mountains. Namely, the vegetation type on this mountains changes with the altitude from alpine steppe in a range 4,850 to 5,200m to subnival belt in a range 5,200 to 5,400m.

East Kunlun Mountains also can be classified into continental semiarid montane type. We can find three vertical vegetation belts: alpine steppe (4,900-5,200m), subnival belt (5,200-5,700m), and nival belt (5,700-6,973m) on the south slope, and three vertical vegetation belts such as alpine steppe (4,200-4,700m), subnival belt (4,700-5,350m), and nival belt (5,350-6,973m) on the north slope.

Altin Mountains belong to continental arid montane type. The vegetation zones on the south slope of this mountains consist of montane shrub desert (3,100-3,400m), montane dwarf suffruticous desert (3,400-3,800), and alpine desert (3,800-4,645m). The vegetation type on the north slope consists similarily of three belts: montane shrub desert (1,000-3,200m), montane dwarf suffrutious desert (3,200-3,600m), and alpine desert (3,600-4,645m).

In these districts, temperature and precipitation decrease gradually with the increment in latitude, which indicates that climatic resources controlling plant growth become less from the southern regions to the northern regions. The reduction in the climatic resources towards the north direction is the most important reason for the simplification of species complexility of vegetation and of regional differentiation of vertical vegetation. The vegetation type in this district changes from marine humid montane type in the southern region, through continental semiarid montane type in the middle region, to continental montane type in the northern region. The difference in the snow line between the south and north slopes of this plateau, and the upper altitude of some vegetation zone distributed vertically on the Hohxil mountains is known to be higher on the south slope than on the north slope.

2.3 Permafrost

The permafrost area on the Tibetan Plateau is reported to be about 1,500,000km^2, which is about 70% of the frozen ground area in China. Most of the frozen ground area in the world is mainly located on the middle latitude of the northern hemisphere. These frozen ground is formed and maintained by severe climate in winter. It is expected that the area of frozen ground, especially the permafrost area would decrease considerably by global climatic warming due to the intensification of the atmospheric greenhouse effect. This is mainly because the frozen ground is very sensitive to temperature change. Therefore the observation of expansion and/or decline of frozen ground area under changing climates could be used as an indicator of changes in global climate. The observation data on the expansion and/or decline of the frozen ground area will help to understand how terrestrial ecosystems will respond to global climatic changes. Color Plate 9 indicates the geographical distribution of the frozen ground on the Tibetan Plateau.

3. MODELLING OF VEGETATION CHANGE

3.1 Holdridge life zone classification system

This system has been widely used to study on a continental scale climate and vegetation (biomass) interaction (e.g. Emanuel *et al.*, 1985). This system classifies natural vegetation formations into thirty types using the three climate indices such as average annual precipitation (*P*, mm), mean annual biotemperature (BT,°C), and potential evapotranspiration rate (PER). BT is the average temperature of monthly temperatures over a period with monthly temperatures above 0℃. PER is the ratio of the potential evapotranspiration (PET) to the precipitation. When this system applies to the classification of natural vegetation on the Tibetan Plateau, most of the alpine zones of this plateau are classified into nival belt. This is because the Holdridge life zone classification system was originally developed for the classification of vegetation formation in tropical climatic zone. Therefore, this system was improved to classify the natural vegetation on the Tibetan Plateau using climatic indexes. The improved system was found to describe and classify reasonably the distribution map (see Color Plate 10) of vegetation on the Tibetan Plateau. Accordingly, this improved classification system can be used to estimate the probable response of the natural vegetation on the Tibetan Plateau to global climate changes.

3.2 Permafrost model

The following air frost number due to Nelson and Outcalt (1987) is often used to quantify the geographical distribution of permafrost.

$$F = \frac{\mathrm{DDF}^{1/2}}{\mathrm{DDF}^{1/2} + \mathrm{DDT}^{1/2}} \tag{2}$$

where DDF and DDT are the freezing and thawing indices (°C day) for a study area, respectively. These two indices can be determined from

$$\left.\begin{aligned} \mathrm{DDF} &= -T_w\, L_w \\ \mathrm{DDT} &= T_s\, L_s \end{aligned}\right\} \tag{3}$$

where T_W, T_S, L_W, and L_S are evaluated from

$$\begin{aligned} T_w &= T - A\left[\sin A/(3.14 - B)\right] \\ T_s &= T + A(\sin B/B), \\ L_w &= 365 - L_s \\ L_s &= 365(B/3.14) \end{aligned}$$

and T, A, and B are given by

$$T = (T_{max} + T_{min})/2,$$
$$A = (T_{max} - T_{min})/2,$$
$$B = \cos^{-1}(-T/A).$$

where T_{max} and T_{min} are, respectively, the air temperatures for the warmest and coldest months. Therefore, T and A define approximately the annual mean air temperature (℃) and the amplitude of annual temperature cycle, respectively. B is the frost angle defined as a point along the time axis at which the annual temperature curve crosses the line of 0℃.

A geographical distribution of frozen ground on the Tibetan Plateau simulated using the above frost number and related weather data is shown in Color Plate 11. The simulated results are quite similar to the frozen ground distribution map based on observation data. As shown in Color Plate 11, no permafrost can be observed in areas with air frost index less than 0.4, and permafrost distributes continuously on areas with air frost index above 0.6. On the other hand, in areas with the air frost index between 0.4 and 0.6, the distribution of permafrost is not continuous and intermittent.

3.3 A model for Net Primary Productivity (NPP) of natural vegetation

The solar energy in dry matters produced photosynthetically by green plants is the fundamental basis for all living things on the Earth. The quantitative determination of energy flow in natural ecosystems has been one of the most important problems relating climate and vegetation. Therefore much research effort has been concentrated on evaluating dry matter production of terrestrial and aquatic vegetations. Three climatic models: Miami model, Thornthwaite Memorial model (Lieth and Box, 1972; Lieth, 1973) and Chikugo model (Uchijima and Seino, 1985, 1988): have been widely used to estimate the magnitude of NPP of terrestrial vegetation from weather data. The first two models are not based on theoretical consideration of climate-plant interaction but on the regression equations obtained from the analysis of data on climate and dry matter production. On the other hand, Chikugo model is based on ecophysiological feature of vegetation and on heat balance analysis. In this model, it is assumed that soil water is enough for plants which are growing luxuriantly. Thus, it is not suitable for semiarid and arid areas (Zhou and Zhang, 1995).

In this report, actual evapotranspiration (AET, mm) was used to build a quantitative model for estimating the magnitude of NPP of natural vegetation on the Tibetan Plateau. As well known, actual evapotranspiration (AET) is one of important predectors of NPP. This is because AET is determined by water availability and solar energy by which the photosynthetic activity of plants is strongly controlled.

The magnitude of AET in a study area is defined as a difference between precipitation and runoff. AET denotes an amount of water evaporated from soil and vegetation over a certain time interval. That is, AET is the sum of evaporation from soil and/or water and transpiration from plants. Therefore, the magnitude of AET depends closely on the availability of water for evaporation and transpiration, solar energy necessary to evaporate water into the air, and air flow for accerelating transfer of water vapor. As Major (1963) pointed out, the magnitude of AET in a study area is quantitatively related to the ecophysiological activity of vascular plants on that area. Rosenzweig (1968) found also that the above ground NPP of terrestrial vegetation has a close correlation with AET. This indicates, therefore, that the determination of AET plays a key role in evaluating NPP of natural vegetation. Although a number of experimental and theoretical studies have been done on the determination of AET, an appropriate and accurate model for this purpose has not been obtained yet.

Using the following two balance equations:

$$P = f + E \quad \text{(for perenial average) and}$$
$$R_n = H + E \quad \text{(for annual average)}$$

Zhou and Zhang (1995) obtained the next model for evaluating the magnitude of AET in a study area.

$$E = \frac{P \cdot R_n \left(P^2 + R_n{}^2 + P \cdot R_n\right)}{\left(P + R_n\right) \cdot \left(P^2 + R_n^2\right)} \tag{4}$$

where P and f are annual precipitation (mm) and annual runoff (mm), respectively, R_n is annual net radiation (mm), E and H denote, respectively, actual evapotranspiration (mm) and sensible heat (mm). By relating Eq.(4) to the data on dry matter production obtained by Efimova (1977), the authors obtained the following model for NPP estimation.

$$\text{NPP} = \frac{P \cdot R_n \left(P^2 + R_n^2 + P \cdot R_n\right)}{\left(P + R_n\right) \cdot \left(P^2 + R_n^2\right)} \exp\left[-\left(9.87 + 6.25\text{RDI}\right)^{0.5}\right] \tag{5}$$

where RDI ($=R_n/P$) is radiative dryness index. NPP-values estimated by this model were well compared with them estimated by the Chikugo model with a good accordance in moist areas. However, our NPP-model was found to give more reasonable NPP-values for arid and semiarid regions than the Chikugo model.

3.4 Ecological Information System

The Laboratory of Quantitative Vegetation Ecology (LQVE) of the Institute of Botany of the Chinese Academy of Sciences developed an Ecological Information System (EIS). This is a kind of Geographic Information System (GIS) and based on the information of ecology, statistics, geosciences, quantitative ecology, and computer technology. In this system (EIS), various databases are used to make clear ecological characteristics of terrestrial vegetation. EIS has been widely used in studies of plant production and biodiversity of Chinese vegetation in relation on global climate changes for long time.

EIS can provide a visualized multivariable analysis program package, a self-defined functional bank, and a four-generation language-Ecological Information Description Language (EIDL). EIS can apply for multidimensional ecological analysis and modelling of climate-vegetation interactions.

The visualized multivariable analysis program package consists of a series of multivariable analysis programs that are used for the data analysis such as multivariable analysis, regression analysis, and principal component analysis, visual demonstration of results so obtained. If a study area is specified and inputed in this package, this package can search accurately various kinds of related ecological data on this study area, make necessary statistical analysis, and demonstrate visually obtained results.

Self-defined functional bank provides the data analysis method in which includes empirical formulae and ecological models. The fundamental structure and operating method of this functional bank are quite similar to those of general databases.

EIDL is a kind of explanatory languages. This can directly use data from various databases and digitalized data, and make linear and/or non-linear analyses. This can easily express and treat the idea, methods, and models of researchers. Therefore, this language can use to build a method bank and expert systems.

The data on EIS can be inquired directly from geographic coordination. EIS spatial data has three points superior to GIS. First point is that any kinds of scatter and graphic data can merge into databases and use together in the analysis according to their coordination. The second point is the automatical zooming of data without any limitation. Any different regions with various scales can be easily manipulated on the computer in which the EIS data management is transplanted. Therefore, worldwide nationwide, and regional data can be used simultaneously. Third, nine projections of results so obtained are always automatically selected according to the necessity of users.

The studies of ecological phenomena with different spatial scales such as community, transect, region, and/or globe can be divided into point, line, and area analyses in EIS. The point analysis is a comprehensive analysis based on a single point coordination. For example, when a study point is selected on a related map using a mouse, the ecogeographical information, such as altitude, meteorological factors, physical and chemical properties of soil, corresponding life zone, and NPP-value for this point and the surrounding 10 points will be obtained on the computer. The point analysis also contains an expert inquest, which consists of deduction expressions and procedures made by EIDL program or functional models. For example, an expert inquest for vegetation classification consists of the following formula

$$LZ = BT + PER \tag{6}$$

where LZ, BT, and PER represent the EIDL programs for the Holdridge life zone classification system, biotemperature, and potential evapotranspiration rate, respectively. If biotemperature and precipitation are given for a study area or point, a vegetation type for the study area is automatically designated on the basis of Holdridge life zone classification systems.

Profile analysis of vegetation distribution is a continuous series of point analysis, because the profile of vegetation distribution can be considered as an arbitrary section defined by the trace of a moving point. The section so defined is composed of different profiles of attributes from database, function bank, and EIDL program. Those indicate complicated relationships between various environmental elements along the section. Color Plate 12 is an example showing results of the profile analysis made by Zhang and Yang (1995) for Northeast China Transect (CENTI). This transect extends from a northern point of Great Xin'an Mountains, through Northeast China mountain regions, North China Plain, Chang Jiang-Hui River Plain, and South China Hilly area, to a southern end point of Hainan island. The points on the above transect are located over the range E109°30' to E128° longitude and N18°44' to 53° latitude. The altitude of respective points was found to be between 0m and 1300m. In Color Plate 12, the X-axis is the geographical coordination and the Y-axis shows meteorological elements and net primary productivity. Color Plate 12 reflects approximately general relationships between various ecological elements and changes in vegetation pattern along the transect.

Area analysis is made according to the following steps. First, the surrounding boundary of a study area is determined. After that, the multivariable analysis method is selected in EIS. Also as the needs, other softwares for statistical analysis, such as SAS and/or MINITAB is loaded from outside systems into our systems. EIS can work on specified data within the given surrounding boundary and give final results as color plates and graphs, in which ecological structure of terrestrial vegetation, dynamic changes in environmental conditions, and spatial distribution of various vegetation types within the study area. The area analysis program developed in our Institute includes also different indicating methods, such as an expression of discreate distribution, curve diagram, bar display, ellipse illustration, and so on.

At present, EIS is being constructed by accumulating the knowledge and information on vegetation and environment through wide application studies. The database, analysis procedures, and system structure for EIS are based on well knitted and standard model structure. EIS is developing gradually with the progress of research work in LQVE and has been successfully used in studies of ecology and environmental sciences. As can be expected from the characteristics of the texture and function of EIS, this can, of course, support research work related to geographical areas.

4. RESPONSES OF NATURAL VEGETATION ON THE TIBETAN PLATEAU TO GLOBAL CLIMATE CHANGE

In order to study probable responses of natural vegetation on the Tibetan Plateau to global climate change, a simple climate scenario equivalent to CO_2 doubling condition was used. That is, annual air temperature and annual precipitation were expected to increase by 4°C and 10%, respectively.

4.1 Response of life zones on the Tibetan Plateau to global climate change

The improved Holdridge life zone classification system and the climate scenario for CO_2 -doubling conditions were used to make clear the response of natural vegetation on the Tibetan Plateau to global climate changes.

Color Plate 13 indicates that the subzones or lifezones of natural vegetation on the Tibetan Plateau will shift northwards from the each present position due to the global climate warming. As Color Plate 13 shows, the shift of each subzone is different between them, reflecting the difference of climatic change among individual regions. Table 2 summarizes the possible changes in the area of each subzone of natural vegetation on the Tibetan Plateau under global climate warming conditions.

Table 2 The possible changes in life subzone areas at present and CO_2-induced global change

	At present (100km^2)		CO_2- induced global change (100km^2)		Changing area (100km^2)
Tropical moist forest	0	(0)	4118	(0.16)	4118
Subtropical moist forest	18723	(0.73)	39375	(1.53)	25602
Warm temperate deciduous broad-leaved forest	33106	(1.29)	146613	(5.70)	113497
Montane coniferous broad-leaved mixed forest	303318	(11.80)	353219	(13.74)	49901
Subalpine coniferous forest	446014	(17.35)	422332	(16.43)	-23682
Montane steppe	164246	(6.39)	427705	(16.64)	263459
Alpine steppe	149743	(5.82)	62228	(2.42)	-87515
Alpine meadow/Tundra	253350	(9.85)	0	(0)	-253350
Montane desert	410730	(15.98)	703162	(27.35)	292432
Alpine desert	287906	(11:20)	319420	(12.42)	31514
Frigid desert	204566	(7.96)	77629	(3.02)	-126937
Subnival	296030	(11.51)	15052	(0.59)	-280978
Nival	3252	(0.13)	141	(0.01)	-3111
Total	2570984	(100%)	2570994	(100%)	

Under the conditions that annual mean temperature and annual precipitation increase, respectively, by 4°C and 10%, in general, the mountain vegetation zones in the southeast Tibetan Plateau will become forests. Namely, the percentage area of mountain forests composing mostly of tropical and temperate forests will reach about 6.4%. On the other hand, the area of alpine meadow will decrease considerably, because montane cold-temperate coniferous forests will replace the most of alpine meadow. Similarily, the area of alpine steppe will become less than 1/2 of the present area, mainly due to that temperate steppe will replace the most of alpine steppe. As shown in Table 2, the alpine desert area in the western part of this plateau will be replaced by temperate desert due to the global climate warming. Table 2 indicates clearly that the area of temperate desert (montane desert) will increase by 12% of the total land area.

This result implies that the speed of desertification occurring on this district at present would be accerelated by the global climate warming.

4.2 Response of montane vertical vegetation belt system to global climate change
Changes in temperature patterns were used in investigating the probable changes in the vertical distribution of natural vegetation due to global climate warming. However, changes in precipitation pattern were disregarded in investigating this problem; the sensitivity test of vertical distribution of vegetation was made against temperature alone. In order to make clear temperature changes on mountain slopes in this district, the following values of the temperature lapse rate (TLR, °C/100m) were adopted for individual slopes, respectively.

0.58°C	on the south Himalayas
0.65°C	on the north Himalayas and south Gangdise Mountains
0.55°C	on the north Gangdise Mountains
0.40°C	on the Hohxil Mountain and southeast Kunlun Mountains
0.71°C	on the northeast Kunlun Mountains and south Altin Mountains
0.67°C	on the north Altin Mountains

Considering the global climate warming (4°C) and the TLR-values mentioned above, it was expected that the upper limit for each vegetation belt on the Tibetan Plateau will rise as follows:

690m	on the south Himalayas
620m	on the north Himalayas and south Gangdise Mountains
730m	on the north Gangdise Mountains
1,000m	on the Hohxil Mountain and southeast Kunlun Mountains
560m	on the northeast Kunlun Mountains and south Altin Mountains
600m	on the north Altin Mountains

For example, reflecting the altitudinal shift of the upper limit of vegetation belts due to global climate warming, montane shrub-grass tussock or shrub-steppe will replace the alpine steppe and nival belts on Hohxil Mountains, and the base belt on alpine steppe belt as the base belt on the northwest Kunlun Mountains is replaced by montane shrub-grass tussock or shrub-steppe. In general, the each montane vegetation belt will shift upwards 560 to 1,000m, reflecting the climate warming. Accordingly, the snow line on mountains will shift upwards, resulting in the disappearance of snow cap on some mountains. As already pointed out in Table 2, the altitudinal shift of vegetation belts on the Tibetan Plateau will cause the spread of montane desert area, implying that a current trend of desertification in this district will be accerelated by the future global climate warming.

4.3 Response of permafrost area to global climate warming

Color Plate 11 shows the spatial distribution of frozen ground area on the Tibetan Plateau to be expected under global climate warming conditions. Under global climate warming conditions, the boundary between continuous and discontinuous (intermittent or seasonal) permafrost areas will shift towards the center area of the plateau about 200km in the east and west parts of the plateau, and more than 200km in the east and west parts of the plateau.

This means evidently that the permafrost area will reduce because of the melting of underground ice by increased temperature. Namely, the area of continuous permafrost will decrease 841,148km^2 at present to 239.922km^2 under global climate warming conditions, indicating that the percentage area of continuous permafrost zone (to the total area of frozen ground on the Tibetan Plateau) will reduce 32.6% to 8.5%. On the other hand, it is expected that the area of the discontinuous permafrost zone will be not very sensitive to global climate warming. Namely, a future area of the discontinuous permafrost zone under global climate warming conditions will be 807,939km^2, while its current area is 879,486km^2.

4.4 Response of net plant productivity (NPP) to global climate change

NPP-values of natural vegetation on the Tibetan Plateau were estimated using the improved NPP-model (Zhou & Zhang, 1995). Although net radiation (R_n) has usually been used in the calculation of net primary productivity (NPP) of natural vegetation, in this study potential evapotranspiration rate (PER) was used to evaluate NPP-values from climatic data. This is mainly because the calculation of R_n needs data on many climatic factors. Using climatic data from 671 weather stations in China, the following empirical formula was obtained between potential evapotranspiration rate (PER) and radiative dryness index (RDI) due to Budyko (1956).

$$\mathrm{RDI} = (0.629 + 0.237\mathrm{PER} - 0.00313\mathrm{PER}^2)^2 \qquad R = 0.95 \tag{7}$$

PER-values can be easily estimated from

$$\mathrm{PER} = 58.93\mathrm{BT} / P \tag{8}$$

where BT and P are biotemperature (°C) and annual precipitation (mm). Using the data on RDI, BT, and P, NPP-values of natural vegetation on the Tibetan Plateau are calculated and the results so obtained are presented in Color Plate 14. As shown in this color plate, net primary productivity (NPP, t dry màttèr / (ha·yr)) increases considerably with the movement from the northwest corner of the Tibetan Plateau to its southeast corner. That is, it increases from nearly null for a severe cold desert region, through about 4 t/(ha·yr) for cold temperate moist forest region, to about 24 t/(ha·yr) for tropical moist forest region.

Color Plate 15 indicates changes of geographical distribution of NPP-values on the Tibetan Plateau under global climate changes. Although it is expected that higher CO_2-concentration in the air may act as a fertilizer to plants promoting the photosynthetic activity of plants, this paper investigated probable effects of global climatic changes alone on NPP-values. Therefore, it is needed to consider fertilizer effects of higher CO_2-concentration on plants in order to improve the present results of vegetation distribution on the Tibetan Plateau.

5. CONCLUSIONS

Tibetan Plateau is thought to be one of the most sensitive or vulunerable areas to climate changes in the world, because Tibetan Plateau has the average altitude of about 5,000m, and its climatic conditions are very severe for the survive of all living things including plants. In this report, probable effects of global climate changes due to CO_2-doubling on the natural vegetation in the Tibetan Plateau were studied using an improved NPP-model, permafrost model, and so on. The results obtained through data analyses can be summarized as follows:

(1) The mountain vegetation consisting of alpine meadow and alpine steppe in the southeast Tibetan Plateau will be replaced by forests. For this reason, the area of those two vegetation zones will reduce with global climate warming. In the western part of the Plateau, temperate desert zone will replace the most of alpine desert zone. Such an exchange of vegetation zones is expected to accerelate the speed of desertification occurring in this district.

(2) Reflecting the global climate warming, the individual upper altitudes of montane vegetation zones (or belts) will shift upwards by 560-1,000m, with the proportional upward shift of the snow line on mountains. Such an upward shift of the snow line will cause a disappearance of snow cap on several mountains.

(3) The boundary between the continuous and discountinuous permafrost regions will shift by about 200km towards a center part of the Tibetan Plateau, resulting in the shrinkage of the permafrost region. The shrinkage of the permafrost region due to global climate warming also accerelate the desertification in this district.

(4) The global climate changes due to CO_2-doubling are expected to cause generally the increment in NPP-values of potential natural vegetation on the Tibetan Plateau. However, the magnitude of NPP increase of the potential natural vegetation would differ among regions depending on the difference of local climatic conditions.

REFERENCES

Budyko, M.I., 1956. *Heat Balance of Earth's Surface.* Hydrometeorological Printing House, Leningrad, 254pp.

Efimova, N.A., 1977. *Radiative Factors of Vegetation Productivity.* Hydrometeorological Printing House, Leningrad, 215pp.

Emanuel, W.R., Shugart, H.H. and Stevenson, M.P., 1985. Climatic change and the broad-scale distribution of terrestrial ecosystem complexes. *Clim. Change,* **1**, 29-43.

Leith, H., 1973. Primary production: Terrestrial ecosystems. *J. Hum. Ecol.*, **1**, 303-332.

Lieth, H. and Box, E., 1972. Evapotranspiration and primary productivity: C.W. Thornthwaite memorial model. *Climatology*, **25**, 37-46.

Major, J., 1963. A climatic index to vascular plant activity. Ecology, **44**, 485-498.

Nelson, F.E. and Outcalt, S.I., 1987. A computational method for prediction and regionalization of permafrost. *Arct. Alp. Res.*, **19**, 279-288.

Rosenzweig, M.L., 1968. Net primary productivity of terrestrial communities: prediction from climatological data. *Am. Natural.*, **102**(923), 67-74.

Uchijima, Z. and Seino, H., 1985. Agroclimatic evaluation of net primary productivity of natural vegetation. (1) Chikugo Model for evaluating primary productivity. *J. Agr. Meteorol.*, **40**, 343-352.

Uchijima, Z. and Seino, H., 1988. An agroclimatic method of estimating net primary productivity of natural vegetation. *JARQ*, **21**, 244-249.

Zhang, X. and Liu, C., 1994. A predictive scenario of vegetation changes on the Tibetan Plateau under global change condition. In: *Global Change and Ecosystems*, (ed. by X. Zhang), 17-26, The Printing House of Shanghai Science and Technology.

Zhang, X. and Yang, D., 1995., Allocation and study on global change transects in China. *Q. Sci.*, 1, 43-52.

Zhou, G. and Zhange, X., 1995. A natural vetetation NPP model. *Acta Phytoecolog. Geobot. Sinica,* **17**(3), 1-8.

EFFECTS OF ELEVATED CO_2 AND GLOBAL CLIMATE CHANGE ON RICE YIELD IN JAPAN

Takeshi Horie[1], Tsutomu Matsui[1], Hiroshi Nakagawa[1] and Kenji Omasa[2]

ABSTRACT: Long-term CO_2 x temperature experiments on rice using Temperature Gradient Chambers (TGCs) revealed that nearly doubled CO_2 concentration in the atmosphere increases crop dry matter production by 24 % through the enhancement of radiation conversion efficiency to biomass, but not through the promotion of radiation interception rate. Although temperature gave a negligibly small influence on the CO_2 effect on biomass production, it significantly affected the grain yield through spikelet fertility. High temperatures were most damaging rice spikelets at the moment of their flowering and made them sterile. The temperature (T_{mh}) at which 50 % spikelets are sterile was 38.2°C for a high temperature-tolerant genotype 'Koshihikari', and 36.6°C for a sensitive genotype 'Akihikari'. Those findings from TGC experiments were parametrized and incorporated into rice growth simulation model SIMRIW.

Impact assessments of elevated CO_2 concentration and global climate change on regional rice yield in Japan were made by using SIMRIW under various climate scenarios. The model predicted that a 100 μmol mol^{-1} increase in CO_2 concentration alone increases rice yield in Japan by 7 - 8 %, and that a 2°C temperature rise significantly reduces this CO_2 effect in the most part of Japan. Then, the effects of 2 x CO_2 climates on regional rice yield were predicted for high temperature-sensitive and tolerant genotypes by using 2 x CO_2 climate scenarios generated by GFDL, GISS and UKMO general circulation models. It was found out that the 1.6°C difference in the spikelet sensitivity to high temperature damage between those two genotypes gives significantly large effects on the predicted rice yield in central-south (Tokai and South Kanto districts) and south-western Japan under 2 x CO_2 climates. Even with the high temperature-tolerant genotype, rice yield in those regions was predicted to decrease by 0 - 10 % than the present under 2 x CO_2 climates except for GISS, while that in northern Japan was predicted to increase by 10 - 25 % depending on climate scenarios. The model also predicted that 2 x CO_2 climates significantly increase yearly yield variability in the central-south and south-western Japan, while they stabilize the yield fluctuation in the northern Japan. Alternations of phenological development genotypes and cropping seasons furthermore promoted the positive effects of 2 x CO_2 climates in the northern Japan, but did not significantly mitigate the negative effects in the southern

[1] Faculty of Agriculture, Kyoto University, Kyoto 606, Japan.
[2] National Institute for Environmental Studies, Tsukuba, Ibaraki 305, Japan.

Japan. For mitigating the predicted negative effects of 2 x CO_2 climates in the southern Japan, explorations for high temperature-resistant genotypes are necessary.

KEY WORDS: rice, elevated CO_2, temperature gradient chamber(TGC), simulation model for rice-weather (SIMRIW), Japan

1. INTRODUCTION

Although Japan relies on imported grains from abroad for two-thirds of the total consumption, its staple food rice is the only crop that it self-supplies. Rice production in Japan is under the control of the government to produce approximately 10 million tons annually which just meet the national demand. This production plan of rice, however, is occasionally disturbed by climatic variations : over production in favorable climate years and shortage in adverse climate years. Except for typhoons which occasionally attack some district of Japan, variations in temperature and solar radiation during rice growing seasons are the major factors that bring yearly variations of Japanese rice production (Yoshino *et al.*, 1988). The cool summer in 1993, for instance, caused 26 percent reduction of Japanese rice production and forced Japan to import more than two million tons of rice, resulting in abrupt highering of rice price in the international market. However, unlike most other countries in the world, water plays a negligible roles as a yield reducing factor in Japanese rice production, because almost all rice fields in Japan are fully irrigated.

Since Japanese rice production is so sensitive to temperature and solar radiation, it is requested to assess the impacts of elevated CO_2 concentration in the atmosphere and anticipated global warming (Hansen *et al.*, 1984) on regional rice production in Japan. To answer this, we have been conducting various experiments on CO_2 x temperature effects on rice by using a Temperature Gradient Chamber (TGC) newly developed (Horie *et al.*, 1991; Kim *et al.*, 1992). The data from TGC experiments were parametrized and incorporated into SIMRIW or SImulation Model for RIce-Weather relationship to simulate rice growth and yield in relation to CO_2 concentration, temperature and solar radiation conditions (Horie, 1993). By applying SIMRIW for Japan with doubled CO_2 climate data predicted by general circulation models for dynamics of earth's atmosphere (GCMs), Horie *et al.* (1995b) predicted that doubled CO_2 concentration and global warming will increase rice yield in northern Japan by 10-30 %, but decrease it in south-western Japan by 10-40 % depending on climate scenarios. Similar results have also been predicted for rice yields in Asia under a doubled CO_2 climate by Matthews *et al.* (1995) by using their rice model ORYZA 1.

These predictions, however, had two major limitions. First, current cultivars and cropping seasons were assumed for predicting rice yield under changed climates, though these are likely to change under doubled CO_2 climates. Second, a value derived from experiments only on 'Akihikari' rice in the TGC was utilized for a crop parameter related to high temperature-induced sterility of rice spikelets, despite this parameter may give a significant effect on the predicted yield under a 2 x CO_2 climate. Furthermore, it was found that a cultivar difference exists in the value of this parameter (Matsui and Horie, 1992). The objectives of this simulation study are three folds. The first is to clarify the effect of the genetical difference in spikelet sensitivity to high temperature damage on regional rice yield in Japan under doubled CO_2 climate. The second is to predict to what extent alternations of cultivars and cropping season can mitigate or promote negative or positive effects of doubled CO_2 climate on regional rice production. On the basis of these analyses, the third is to predict most probable effects of doubled CO_2 concentration and anticipated global climate change on regional rice yield in Japan.

This paper briefly describes the model SIMRIW, experimental derivations of its parameters, and the model performance, and then represents the simulation results on the effects of elevated CO_2 and predicted climate change on regional rice yields in Japan.

2. BASIC STRUCTURE OF THE MODEL SIMRIW

Since the details of SIMRIW, the rice model applied for this study have already been reported (Horie, 1987; Horie *et al.*, 1992, 1995a), only a brief description is made here about the basic structure of SIMRIW.

SIMRIW consists of three major parts that describe the processes of ontogenetic crop development, biomass accumulation and yield formation. The ontogenetic development process of rice from emergence to heading is represented in SIMRIW by a continuous variable, the developmental index (DVI), the value of which is defined to be zero at emergence, 1.0 at heading and 2.0 at maturity. The value of DVI at any given moment of crop development is given by integrating the developmental rate (DVR) with respect to time. The DVR is given by a nonlinear function of daily mean temperature and day length as described in Horie and Nakagawa (1990).

The dry matter accumulation process of the rice crop is simulated using the idea that crop dry weight at any moment is proportional to the absorbed solar radiation accumulated up to that moment (Monteith, 1977). This process of biomass accumulation is characterized by only one crop parameter, the solar radiation conversion efficiency (*Cs*). The *Cs* is assumed to be constant up to heading (DVI=1), and thereafter is given by a decreasing function with respect to DVI to simulate the maturation or senescence process. The leaf area growth rate which governs the radiation absorption rate is modeled as a unique function of temperature. This is in contrast to the traditional approach in which leaf area growth is calculated from its weight growth through multiplying by a simple conversion factor, the specific leaf area. The leaf area growth is modeled independently of its weight growth in this model, by taking into account the largely independent nature of each to the other, as described by Horie *et al.* (1979).

The grain yield is simulated in SIMRIW from calculated total biomass by multiplying by harvest index. Harvest index is given by a function of DVI and the sterility percentage of spikelets. The harvest index-DVI relationship is employed to make the yield formation process dynamic and to simulate premature cessation of growth when crops encounter autumn coolness. The sterility percentage of spikelets is given by a function of the cooling degree-days (Uchijima, 1976) during sensitive period of spikelets to cool temperature (0.75<DVI<1.2).

SIMRIW gives a climatically potential yield of a given cultivar of irrigated rice under optimal cultivation technologies. Horie (1987) showed that a close linear relationship existed between the simulated potential yield (Yp) and actual yield (Ya) across respective locations in Japan and in the USA. The relationship could be represented by,

$$Ya=KYp \tag{1}$$

in which K is regarded as an index of the overall technology level of rice cultivation converted into the farmers' yield at the present day.

3. EXPERIMENT AND PARAMETRIZATION ON CO_2 AND HIGH TEMPERATURE EFFECTS ON RICE

For the application of SIMRIW for impact prediction of doubled CO_2 climate on rice yield, parametrizations of the effects of CO_2 concentration and high temperature on rice are indispensable. However, information about this effect based on long-term CO_2 x temperature treatments on rice is very limited to those of Imai *et al.* (1985) and Baker *et al.* (1990a, b). For this reason, we conducted long-term CO_2 x temperature treatment experiments on rice, using a newly developed temperature gradient chamber (TGC) (Horie *et al.*, 1995c). The TGC is a 25 m-long plastic-covered tunnel in which the air flows from one end to the other at a given rate to create a temperature gradient along its longitudinal axis by solar energy in daytime (Fig.1). At night time, the direction of the air was reversed, and heated air by an oil-heater at the daytime exhaust end was led into TGC to create a temperature gradient along the long-axis of TGC due to natural cooling. Two TGCs, one with and one without CO_2 enrichment capability were constructed. Rates of airflow and CO_2 release were controlled by a computer to maintain the temperature difference between the two ends of TGC at 5°C and CO_2 concentration in TGC at 700 umol mol^{-1}. This system enabled long-term CO_2 x temperature treatments on rice under field-like conditions.

A result of a long-term CO_2 x temperature experiment in the TGC is represented in Fig.2 in which dry weights of crop and panicle of 'Akihikari' rice grown under ambient (350 μmol mol^{-1}) and elevated CO_2 (700 μmol mol^{-1}) concentrations are given as a function of average temperature over the growth period. The doubled CO_2 concentration increased total dry weight of rice by 25 % on average over the entire temperature range (27-32°C). No consistent temperature effect could be seen on the total dry weight in the temperature range of this experiment. However, temperature gave a significant effect on the panicle dry weight both under the elevated and ambient CO_2 conditions : the panicle dry weight decreased sharply as the average temperature exceeded 29°C irrespective of CO_2 concentrations. This sharp decline in the panicle weight at the higher temperature was due to high temperature-induced sterility of rice spikelets. A similar result was also obtained

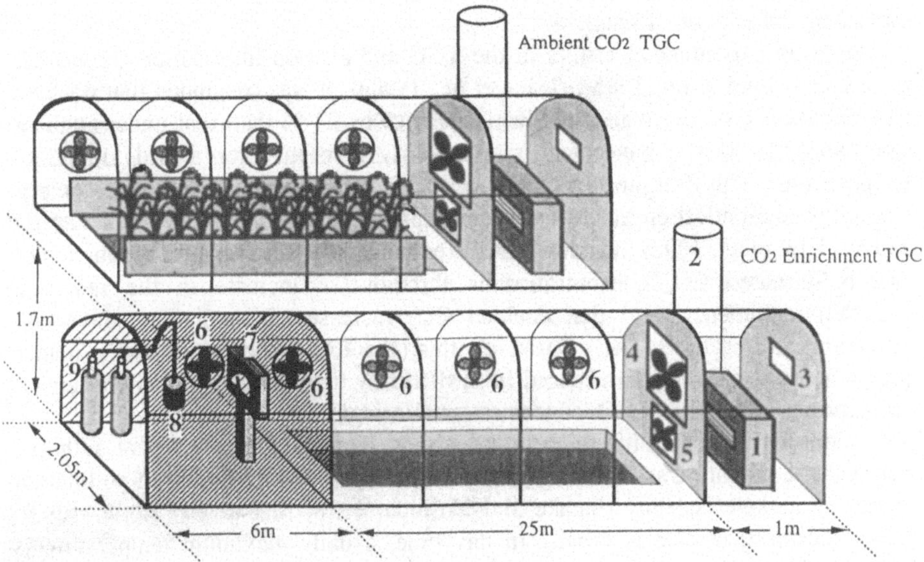

Fig. 1. Schematic drawing of Temperature Gradient Chamber (TGC) of CO_2 enrichment type (bottom) and ambient CO_2 type (top) for the study of temperature x CO_2 effect on rice (Horie *et al.*, 1995c). The numbers on the figure show, 1 : oil heater, 2 : stovepipe of oil heater, 3 : air exhaust window, 4 : variable speed exhaust fan, 5 : reversible exhaust fan, 6 : oscillating fan, 7 : CO_2 controller, 8 : CO_2 injection pipe, and 9 : liquid CO_2 tanks, respectively.

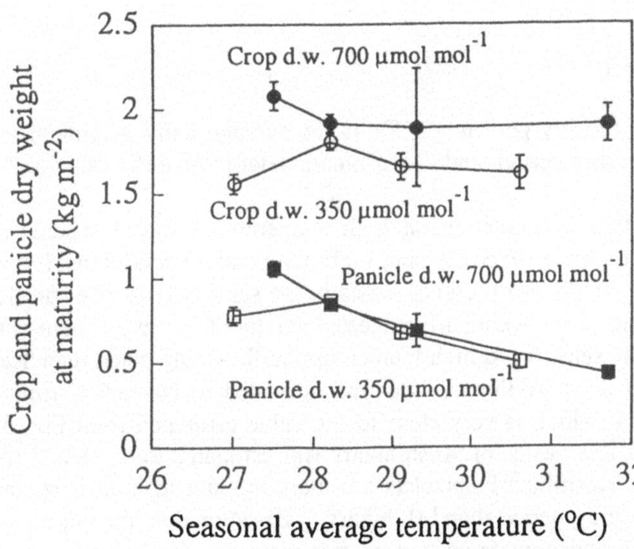

Fig. 2. Total crop and panicle dry weights of rice at maturity grown under elevated (700 μmol mol^{-1}) and ambient (350 μmol mol^{-1}) CO_2 concentrations in the TGC as a function of seasonal mean temperature (Horie *et al.*, 1995c).

for rice by Baker *et al.* (1990a).

Based on experimental results in the TGC and also on information from other researchers (Imai *et al.*, 1985 ; *Baker et al.*, 1990a), it was concluded that doubled CO_2 enhances crop dry matter production of rice by 24 % than that under ambient CO_2, and that this enhancement rate is fairly constant over a wide range of temperature. This finding together with the fact that doubled CO_2 gives only a negligibly small effect on rice leaf area development (Imai *et al.*, 1985 ; Baker *et al.*, 1990a ; Kim *et al.*, 1992) indicates that the enhancement of dry matter production of rice by elevated CO_2 concentration is through the increase in the radiation conversion efficiency, and that doubled CO_2 increases the radiation conversion efficiency of rice by 24 %. Based on this, the CO_2 effect on rice dry matter production was simply parametrized in SIMRIW by giving the radiation conversion efficiency as a function of CO_2 concentration (Horie, 1993).

Among rice organs, high temperature is most damaging to spikelets at anthesis. When the temperature at flowering period exceeds about 35°C, failure in pollination results in spikelet sterility (Satake and Yoshida, 1978; Matsui and Horie, 1992). Since flowering of rice is usually in day time, a daily maximum temperature is considered to be more closely related to high temperature-induced spikelet sterility of rice than the average temperature. Fertility percentage (100-sterility percentage) of spikelets was plotted against average daily maximum temperature over 10 day-period around the heading stage for 'Akihikari' rice grown under the elevated and ambient CO_2 conditions in TGC (Fig. 3). The results indicate that CO_2 concentration has no effect on the temperature and fertility relationship. The relation was approximated by (Horie, 1993)

$$F = \frac{100}{1 + \exp\{0.853(T_m - T_{mh})\}} \tag{2}$$

where F is the fertility percentage, T_m is the average daily maximum temperature during the flowering period, and T_{mh} is a parameter to give the value of T_m at which F becomes 50 %.

From a further investigation on high temperature-induced spikelet sterility by using a phytotron under fixed day and night temperature conditions, it revealed that a relatively large cultivar difference exists in the sensitivity of rice spikelets to high temperature (Fig. 4). 'Akihikari' subjected for the TGC experiment (Fig. 3) was found to be more sensitive to high temperature at flowering peiod than 'Koshihikari'. The value of T_{mh} of 'Akihikari' rice was estimated to be 36.6°C from the result shown in Fig. 4, which is very close to the value estimated from Fig. 3. On the other hand, the T_{mh} value of 'Koshihikari' was estimated to be 38.2°C from Fig. 4, indicating that 'Koshihikari' spikelets are more tolerant to high temperature stress during the flowering period than 'Akihikari'. Based on this, the impact assessments of elevated CO_2 and climate change on rice yield by SIMRIW were made for two genotypes of high temperature-sensitive (T_{mh} = 36.6°C) and tolerant one (T_{mh} = 38.2°C).

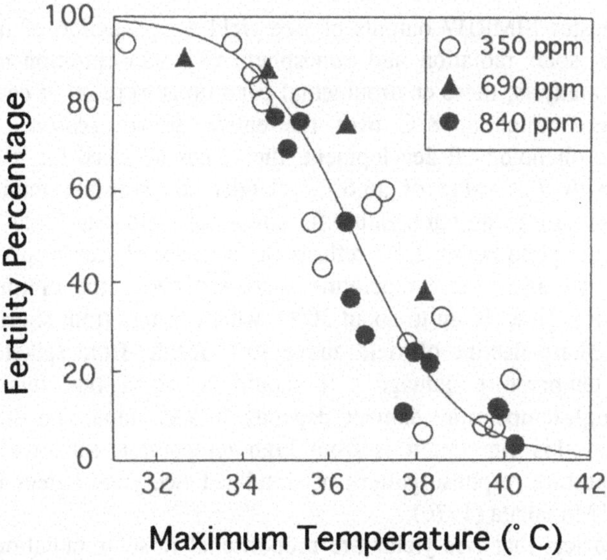

Fig. 3. Relation between average daily maximum temperature during the flowering period and spikelet fertility in the variety Akihikari acclimated to different CO_2 concentrations (Horie, 1993).

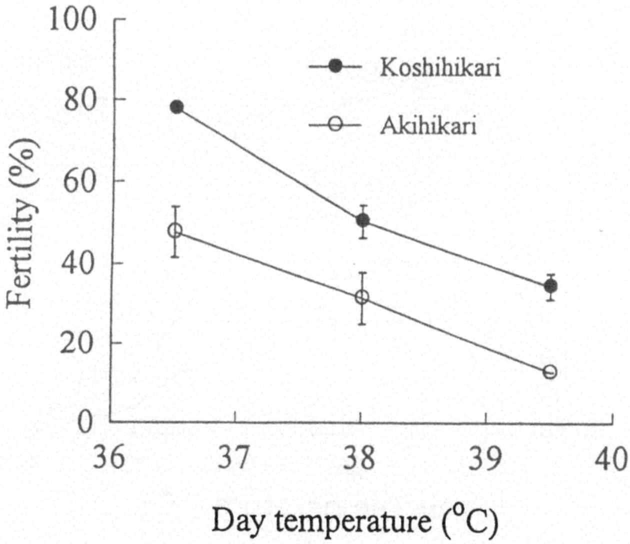

Fig. 4. Spikelet fertility percentages of rice cultivar 'Koshihikari' and 'Akihikari' as affected by day temperature during the flowering period (Matsui *et al.*, unpublished). Plants were subjected to constant day and night temperatures during flowering period.

4. MODEL PERFORMANCE

Figure 5 demonstrates SIMRIW outputs of rice yield as a function of input daily mean temperature, solar radiation and atmosphere CO_2 concentration conditions. In this sensitivity analysis, those environmental conditions were set at constant and diurnal temperature range at 8°C over the entire growth season. As crop parameters for the phenological development, the values obtained for 'Nipponbare' rice were used with *Tmh* value of 36.6°C. Under the constant environmental conditions, the optimum mean temperature for simulated yield was 22-23°C. The sharp decrease of the yield below 22°C reflects the increase of sterile spikelets from cool temperature damage. As temperature increased above the optimum, yield declined more or less linearly up to about 30°C, which results from shortened total growth duration. Sharp decline of yield above 30°C results from spikelet sterility caused by high temperature damage. It should be noted that this threshold temperature for high-temperature damage depends on *Tmh* value, and that the *Tmh* value employed in this simulation is from high temperature-sensitive genotype. The overall temperature response pattern of simulated rice yield agrees fairly well with the results of Munakata (1976).

The model predicts that rice yields are proportional to solar radiation over the entire temperature range and that doubled CO_2 in the atmosphere alone increases rice yield by 24 % under each temperature and radiation condition. These responses resulted from the model hypotheses described in the foregoing sections.

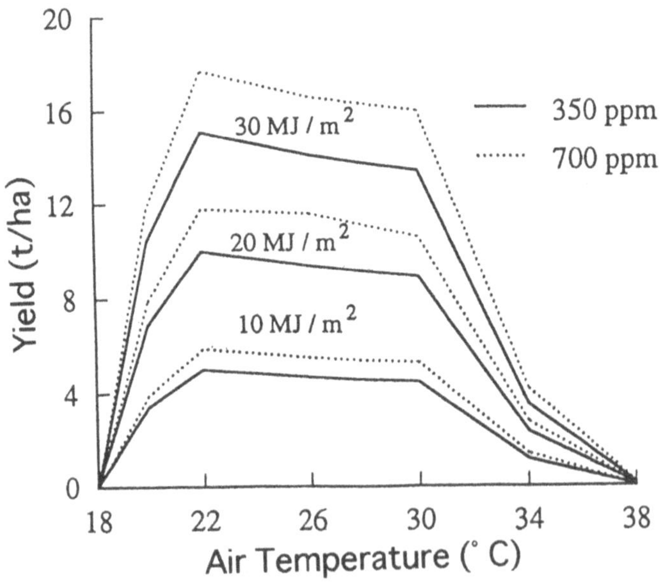

Fig. 5. Simulated yield responses of the cultivar Nipponbare to daily mean temperature, solar radiation, and CO_2 concentration, under constant environmental conditions (Horie *et al.*, 1995a). Day length and diurnal temperature range were set at 12 h and 8°C in each case, respectively.

Validations of SIMRIW were made by using rice growth and yield data obtained at International Rice Research Institute in Philippines and regional farmers' yield data over Japan. Those validations indicated that the model satisfactorily explained difference in rice yield among locations and cropping seasons based on difference in environments (Horie *et al.*, 1995a, b).

5. EFFECTS OF ELEVATED CO_2 AND CLIMATE CHANGE ON REGIONAL RICE YIELD

5.1 Input data

For the assessment of impact of global environment change (elevated CO_2 and climate change) on Japanese rice production, Japan islands were divided into fourteen agro-ecological zones (AEZs) according to Ozawa (1962). Daily weather data for the twelve years from 1979 to 1990 in nine weather stations were selected. The covering AEZs of those weather stations are given in Table 1. Since weather data were not specified for AEZs X and XIII, the climate change effects on rice yield in those AEZs were estimated by interpolation of the predicted yields in the surrounding AEZs.

Table 1 Weather stations, their representing agro-ecological zones (AEZs), and cultivars adopted for the prediction of rice yield under global environment change scenarios in each AEZ.

Site ID	Weather station	Latitude	Longitude	Representing AEZs	Cultivars adopted
1	Sapporo	43° 03'N	141° 20'E	I,II,III,IV	Ishikari
2	Akita	39° 43'N	140° 06'E	VI	Sasanishiki
3	Sendai	38° 16'N	140° 54'E	V	Sasanishiki
4	Maebashi	36° 24'N	139° 04'E	VII,VIII	Koshihikari
5	Toyama	36° 42'N	137° 12'E	IX	Koshihikari
6	Nagoya	35° 10'N	136° 58'E	XI	Nipponbare
7	Hiroshima	34° 22'N	132° 26'E	XII	Nipponbare
8	Kohchi	33° 33'N	133° 32'E	XIV	Nipponbare
9	Miyazaki	31° 55'N	131° 25'E	XIV	Mizuho

From the twelve-year weather data, the 'average' climate for each location was synthesized by averaging daily weather values over the twelve years for each location. This average climate was then used as the baseline to evaluate the effects of climate change on rice yields. Five future climate scenarios were examined : 450 μmol mol^{-1} CO_2 concentration with no change in climate, 450 μmol mol^{-1} CO_2 with a 2°C temperature rise and the 2 x CO_2 climate predicted by the GFDL, GISS and UKMO General Circulation Models (GCMs) of the earth's atmosphere. Those GCMs predicted climate data were supplied by the Data Support Section within the Scientific Computing Division of the National Center for Atmospheric Research through US EPA (Bachelet *et al.*, 1995). The future climate conditions were created by adding the monthly temperature changes in each scenario to the current daily maximum and minimum temperatures of the same month, and by multiplying relative changes in monthly solar radiation by current daily solar radiation values.

As crop parameters for phenological development of rice, the values derived from cultivars listed up in Table 1 were used for each AEZ. Although some of those cultivars are no longer the predominant cultivars in some AEZs at present, those cultivar were adopted for the simulation by assuming that phenological development traits of the present predominant cultivars in each AEZ are not very different from those. Rice yields under the various climate scenarios were simulated in each AEZ for two cases of the crop parameter (T_{mh}) that represents the spikelet sensitivity to high temperature damage at flowering : T_{mh} = 36.6°C for sensitive and 38.2°C for tolerant genotypes to the high temperature damage.

The potential rice yields that SIMRIW simulated were converted into actual farmers' yield at the present day, by multiplying the simulated yields by the technological coefficient (*K* in Eq. (1)). Since regional farmers' yield in most AEZs in Japan are increasing owing to technological advancements, the coefficient *K* in each region is a function of time. Based on multiple regression analysis between the reported actual yield and simulated potential yield for twelve year period in each region, the technological coefficient (*K*) at the present day was determined for each region (Horie *et al.*, 1995b).

5.2 Climate change effects on regional rice yield for high temperature-sensitive and tolerant genotypes

Table 2 shows the predicted change in rice yield under each climate scenario from that of the base (current average) climate for high temperature-sensitive and tolerant

Table 2. Predicted change in potential rice yields from the current values at various locations under different CO_2 and climate conditions.

Climate		Sapporo	Akita	Sendai	Maebashi	Toyama	Nagoya	Hiroshima	Kohchi	Miyazaki
CO_2		**High Temperature Sensitive Genotype**								
ppm										
450	+0℃	+6.7%	+8.0	+7.8	+7.7	+7.7	+7.7	+7.7	+7.7	+7.8
450	+2℃	+0.7%	+6.2	+3.8	+5.2	+2.4	-6.8	+1.9	+3.1	+7.3
640	GFDL	-2.8%	+12.5	+7.3	+7.1	-10.5	-33.9	-21.1	-15.3	+3.7
640	GISS	+13.3%	+26.3	+19.9	+22.3	+18.5	-2.2	-17.4	-29.0	+1.2
640	UKMO	+8.4%	+24.0	+17.7	+13.5	+13.6	-40.9	-47.1	-36.7	-10.0
		High Temperature Tolerant Genotype								
ppm										
450	+0℃	+6.7%	+8.0	+7.8	+7.7	+7.7	+7.7	+7.7	+7.7	+7.8
450	+2℃	+0.8%	+6.2	+3.8	+5.3	+3.0	+4.5	+3.9	+5.7	+7.3
640	GFDL	-2.8%	+12.4	+7.3	+8.5	-9.4	-9.1	-5.1	+0.2	+6.0
640	GISS	+13.3%	+26.3	+19.8	+21.9	+18.5	+18.8	+9.9	+12.5	+22.0
640	UKMO	+8.4%	+24.0	+17.7	+23.2	+13.5	-1.2	-7.8	-2.1	+13.2

genotypes in the nine locations investigated. The results shown in Table 2 were obtained by assuming current cultivars and cropping seasons for all climate conditions. Irrespective of rice genotypes, the model predicted that a 100 μmol mol^{-1} increase in CO_2 concentration over the present level will increase rice yield by 7-8 % at all locations. In conditions where a 2°C temperature rise was added to the 100 μmol mol^{-1} CO_2 elevation, the positive effect of the elevated CO_2 decreased in all the locations, with largest reductions for high temperature-sensitive genotype in southern Japan. The negative effects of the 2°C temperature rise resulted from the facts that higher temperature accelerates rice phenological development and shortens growth duration, and, in southern Japan, causes spikelet sterility for the high temperature-sensitive genotype.

The predicted relative changes of rice yield for the respective locations under 640 μmol mol^{-1} CO_2 and GFDL, GISS and UKMO climate conditions were plotted on the map for high temperature-sensitive (Fig. 6) and tolerant (Fig. 7) genotypes. The predicted effects of 2 x CO_2 climates on rice yield in Japan depend significantly on climate scenarios, adopted rice genotypes and regions. Positive effects of 2 x CO_2 climates, ranging from ca. 10 % (GFDL) to ca. 25 % (GISS and UKMO) yield increase, were predicted in northern Japan, irrespective of rice genotypes adopted. The predicted effects of 2 x CO_2 climates on rice yield in the central-south(AEZ XI) and south-western Japan strongly depend on adopted rice genotypes. For the high temperature sensitive genotypes, severe yield reductions by ca. 10 % (GFDL), 20 % (GISS) and ca. 40 % (UKMO) from the present were predicted in those regions. When the high temperature-tolerant genotype was adopted, those negative effects of 2 x CO_2 climates in the central-south (AEZ XI) and south-western Japan decreased to less than 10 % (GFDL and UKMO) or were reversed into positive (GISS). Note that the temperature (T_{mh}) at which 50 % of spikelets are sterile differed only 1.6°C between the two genotypes. It turned out that the small difference in the spikelet sensitivity to high temperature damage gives significantly large effects on the predicted Japanese rice production under 2 x CO_2 climates.

Since it is very likely that, under a 2 x CO_2 climate, rice cultivars in southern Japan will be endowed with tolerance to high temperature damage of spikelets at flowering, the prediction for the high temperature-tolerant genotype would be a better estimate of probable effects of future climates. Based on this, the predicted rice yield change in Japan under each 2 x CO_2 climate scenario may be summarized as follows : doubled CO_2 with GISS climate will increase rice yield in most regions in Japan by 10 - 25 % ; that with GFDL climate will increase the yield by approximately 10 % in the northern Japan and reduce it by 5 - 10 % in the southern Japan ; that with UKMO climate will increase the yield by approximately 20 % in the northern Japan and reduce it by 5 % in the southern Japan. The largest positive effect in GISS scenario resulted from the fact that GISS model predicts the smallest temperature rise and an increase of solar radiation under doubled CO_2 conditions in Japan, while the smallest positive effect in GFDL scenario resulted from GFDL model predicting a reduction in solar radiation in most part of Japan.

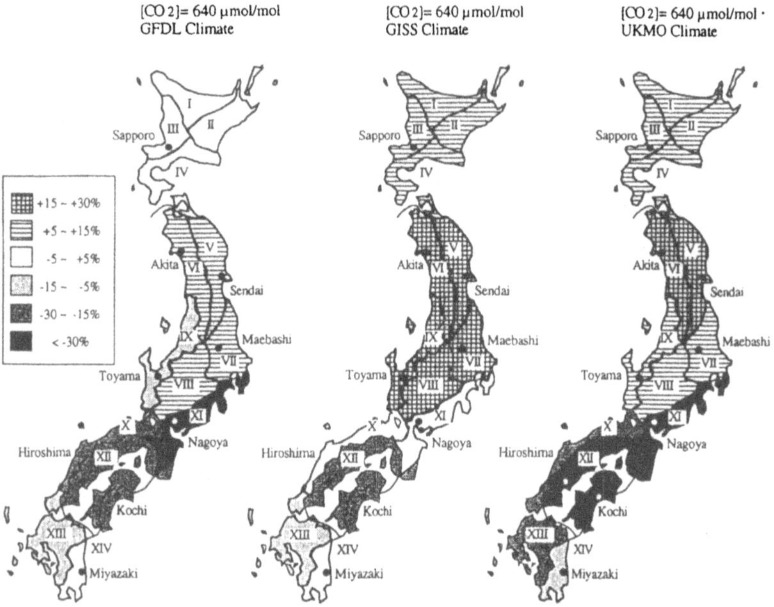

Fig. 6. Effects of 2 x CO_2 and associated climate change predicted by three GCMs on regional rice yield in Japan (prediction by SIMRIW for high temperature-sensitive genotype) (adapted from Horie *et al.*, 1995b).

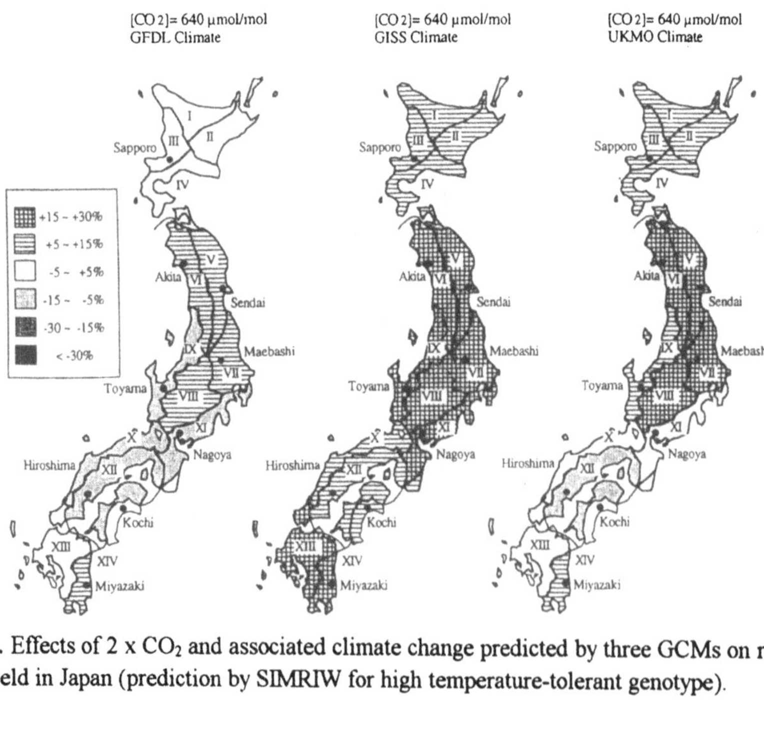

Fig. 7. Effects of 2 x CO_2 and associated climate change predicted by three GCMs on regional rice yield in Japan (prediction by SIMRIW for high temperature-tolerant genotype).

5.3 Probability analysis on effects of GCM climates on rice yield

Probability analyses were made on effects of doubled CO_2 and climate change on rice yield for representative locations, by using daily weather data in the 1979 - 1990 period at Sapporo (AEZ I to IV), Sendai (AEZ V), Nagoya (AEZ XI) and Miyazaki (AEZ XIV) as base climates, and climate change scenarios by GFDL, GISS and UKMO GCMs. Rice genotype with tolerance to high temperature damage of spikelets was adopted for the probability analysis at each location. Figure 8 gives the results of the probability analysis by SIMRIW where the simulated rice yield in each of twelve years under the three different climate scenarios was plotted as cumulative probability distributions.

At Sapporo in Hokkaido (AEZ I to IV), the average predicted yield under the current climate was 5.27 t/ha with a coefficient of variation (CV) of 9.7 %. Yield increases of +4.6 %, + 21.5 % and + 17.5 % were predicted for the 2 x CO_2 climate of the GFDL, GISS and UKMO, respectively. The GISS gave the largest yields of the three GCMs due to the smallest temperature rise and to increased solar radiation levels. Since cool summer damages of rice, which cause a large year-to-year variation of rice yield in Hokkaido, hardly occurred under the 2 x CO_2 climates, significant reductions of the yield variability were predicted there.

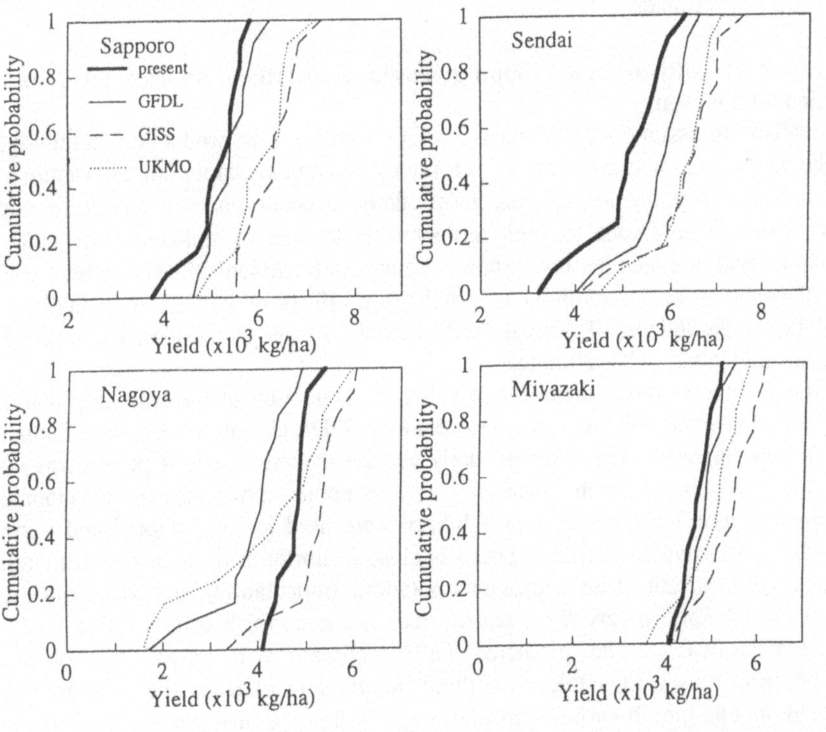

Fig. 8. Cumulative distribution functions for rice yield at four representative locations in Japan under 2 x CO_2 and three scenarios of global climate change (prediction by SIMRIW for high temperature-tolerant genotype).

The largest positive effect of 2 x CO_2 climates was predicted in Tohoku district (AEZ V and VI). At Sendai, the 2 x CO_2 climates not only increased the average yield by 12.5 % to 24.3 %, depending on GCMs, it also reduced the yield variability from the current 15 % to less than 11%.

The largest negative effect of a 2 x CO_2 climate was predicted at Nagoya in Aichi (AEZ XI), where changes in the average yield were - 16.6 %, + 10.4 % and - 9.9 % for GFDL, GISS and UKMO climates. The yield variability was also predicted to significantly increase from the current CV of 7 % to between 11 % and 32 %. These negative effects are due to Nagoya having the highest daily maximum temperature during current conditions of all the locations investigated, so that, further warming in years with hot summer increases the possibility of high temperature-induced spikelet sterility.

At Miyazaki in Miyazaki prefecture (AEZ XIV), between + 3.8 % and + 15.8 % increase of average yield was predicted. However, yield variability was predicted to slightly increase there under 2 x CO_2 climates. The average yield and its variability predicted for Nagoya and Miyazaki are not so catastrophic as those reported in the previous report (Horie *et al.*, 1995b). This is due to the previous report employing the high temperature-sensitive genotype for the prediction, while the present analysis employing the tolerant genotype. Again, it shows that only 1.6°C difference in the T_{mh} value of cultivar gives significant influence on the predicted rice yield in Japan under 2 x CO_2 climates.

5.4 Effect of cultivar and cropping season alternations on rice yield under doubled CO_2 climate

It is realistic to assume that, under a 2 x CO_2 climate, rice production technologies will be adapted to changed climate. Of the technological adaptation, alternations of cultivars and cropping seasons are most likely to occur under a climate change. Apart from the tolerance to high temperature damage of spikelets, rice cultivar adaptation will be made mainly through changes in phenological development traits. The objective of this section is to predict the effects of change in phenological development parameters of cultivars and in cropping seasons on rice yield at selected locations under a 2 x CO_2 climate.

Figure 9 shows the simulated rice yields as a function of transplanting date for current and adapted cultivars under current and GFDL climate condition at Sapporo, Sendai and Nagoya. The 'average' daily climate over twelve-year period was used for this analysis at each location. At Sapporo, phenological development parameters from 'Koshihikari' and 'Ishikari' were used for the adapted and current cultivars, respectively. Since Sapporo is close to northern limit of rice cultivation in Japan, its current climate provides potential transplanting period of only one month for 'Ishikari', a very short season rice, and none for 'Koshihikari', a medium maturing cultivar. The predicted GFDL climate will extend the potential transplanting season of 'Ishikari' to three month, and increase the yield by about 20 % by its one-month earlier transplanting. When the medium maturing cultivar 'Koshihikari' is adopted with one month earlier transplanting under GFDL climate, a 50 % increase in potential rice yield is predicted by SIMRIW. Similarly it is predicted for Sendai that GFDL climate will enable a half month earlier transplanting of 'Nipponbare', a later maturing cultivar than the current 'Sasanishiki', with 13 % increase in the potential yield.

At Nagoya where a negative effect of GFDL climate was predicted, it is predicted that alternation of cropping season will not mitigate the negative effect, despite the climate change significantly extends the cropping season. Replacement of the current cultivar 'Nipponbare' with 'IR 36', an indica type rice, will increase the yield both under the current and GFDL climate conditions, but not mitigate the negative effect of GFDL climate on rice yield there.

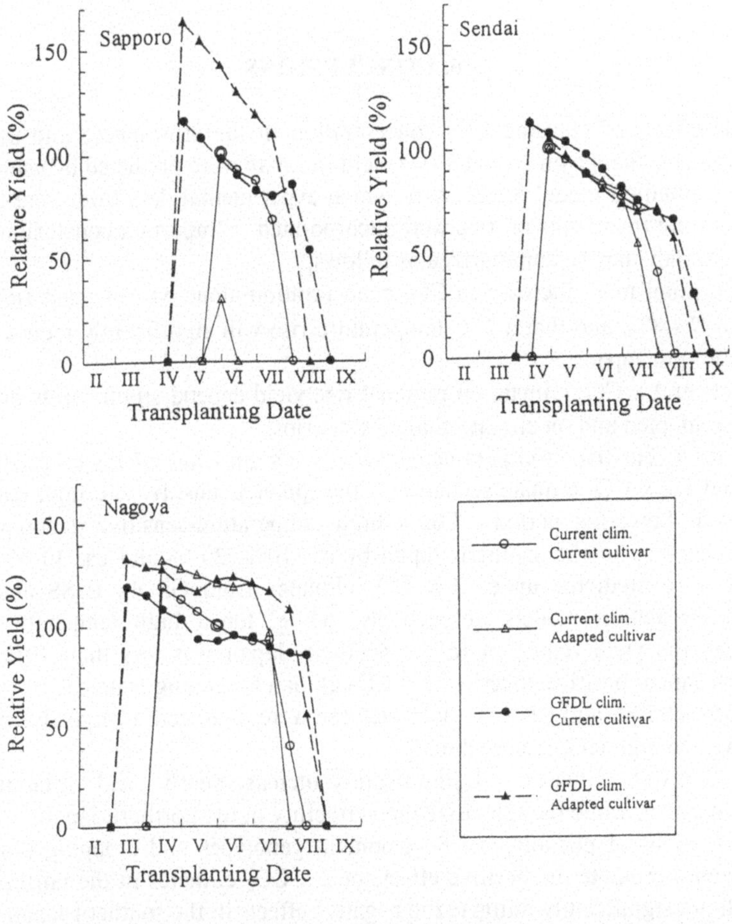

Fig. 9. Relative yield change as a function of transplanting date for current and phenologically adapted rice cultivars under current and GFDL 2 x CO_2 climate conditions for three representative locations (prediction by SIMRIW). The current cultivars used for simulations were 'Ishikari', 'Sasanishiki' and 'Nipponbare' for Sapporo, Sendai and Nagoya, respectively, and the phenologically adapted cultivars were 'Koshihikari', 'Nipponbare' and 'IR 36' for Sapporo, Sendai and Nagoya, respectively. All the cultivars used were assumed to have the same tolerance to high temperature damage of spikelets at flowering as that of 'Koshihikari'. The double circle on the figure denotes the current transplanting date.

The above analysis suggests that adoption of adapted cultivars in phenology and alternation of cropping seasons will furthermore increase rice yield in those regions where a positive effect of 2 x CO_2 is predicted for current cultivation conditions, but will not significantly mitigate rice yield in those regions where negative effects are predicted. In hotter regions where a negative effect of 2 x CO_2 climate is predicted on rice, the only way to mitigate it would be adoption of much resistant genotypes to high temperature-induced spikelet sterility, either with higher tolerance to high temperature at flowering time or with capability to flower early in the morning before temperature rise as reported by Imaki *et al.* (1987).

6. CONCLUSIONS

Probable effects of elevated CO_2 concentration in the atmosphere and anticipated global climate change on regional rice yield in Japan were predicted by using the rice growth simulation model SIMRIW in which experimental data from long-term CO_2 x temperature treatments on rice were incorporated. Important conclusions derived from this study may be summarized as follows:

1. A 100 $\mu mol\ mol^{-1}$ increase in CO_2 concentration alone will increase rice yield in Japan by 7 - 8%, and that a 2° C temperature rise will significantly reduce this CO_2 effects in most Japan.
2. Effects of 2 x CO_2 climate on regional rice yield depend significantly on regions, cultivars adopted and on climate change scenarios.
3. The most sensitive crop parameter that gives significant effects on predicted rice yield under 2 x CO_2 climate scenarios is the spikelet sensitivity to high temperature damage at flowering period. For a high temperature-sensitive genotype, severe yield reductions in the southern Japan by ca. 10%, 20 % and ca. 40 % from the present were predicted under 2 x CO_2 climates from GFDL, GISS and UKMO general circulation models, respectively, while, for a high temperature-tolerant genotype, the yield reduction in the southern Japan was less than 10%. In the northern Japan, positive effects of 2 x CO_2 climates, ranging from ca. 10 % (GFDL) to ca. 25 % (GISS and UKMO) yield increase were predicted, irrespective of cultivar sensitivity to high temperature damage.
4. The 2 x CO_2 climates will significantly increase yearly yield variability in the southern Japan, while they stabilize the variability in the northern Japan.
5. Alternations of phenological development genotypes and cropping seasons will furthermore promote the positive effects of 2 x CO_2 climates in the northern Japan, but will not significantly mitigate the negative effects in the southern Japan.
6. For mitigating the predicted negative effects of 2 x CO_2 climates in the southern Japan, explorations for more resistant rice genotypes to the high temperature damage at flowering are necessary.

REFERENCES

Bachelet, D., King, G.A. and Chaney J., 1995. Climate change scenarios. In: *Modelling the Impact of Climate Change on Rice Production in Asia.* (ed by

R.B. Matthews, M.J. Kropff, D. Bachelet and H.H. van Laar), 67-83, Cab International, Oxon, UK.

Baker, J.T., Allen, L.H., Jr., Boote, K.J., Rowland-Bamford, A.J., Waschmann, R.S., Jones, J.W., Jones, P.H. and Bowes, G., 1990a. Temperature effects on rice at elevated CO_2 concentration. *1989 Progress Report of Response of Vegetation to Carbon Dioxide, No.060.* Plant Stress and Protection Research Unit, USDA-ARS, University of Florida, Gainesville, 70 pp.

Baker, J.T., Allen, L.H., Jr. and Boote, K.J., 1990b. Growth and yield responses of rice to carbon dioxide concentration. *J. Agr. Sci.*, **115**, 313-320.

Hansen, J., Lacis, A., Rind, D., Russell, G., Stone, P., Fung, I., Ruedy, R. and Lerner, J., 1984. Climate sensitivity: analysis of feedback mechanisms. In: *Climate Processes and Climate Sensitivity* (ed by J. Hansen and T. Takahashi), Maurice Ewing Series, 5, 130-163. American Geophysical Union, Washington, D.C.

Horie, T., 1987. A model for evaluating climatic productivity and water balance of irrigated rice and its application to Southeast Asia. *Southeast Asian Studies,* **25**, 62-74.

Horie, T., 1993. Predicting the effects of climatic variation and effect of CO_2 on rice yield in Japan. *J. Agr. Meteorol.*, **48**, 567-574.

Horie, T. and Nakagawa, H., 1990. Modeling and prediction of development process in rice. I . Structure and method of parameter estimation of a model for simulating developmental process toward heading. *Jpn. J. Crop Sci.*, **59**, 687-695.

Horie, T., Nakagawa, H., Centeno, H.G.S. and Kropff, M.J., 1995a. The rice crop simulation model SIMRIW and its testing. In: *Modelling the Impact of Climate Change on Rice in Asia* (ed by R.B. Matthews, M.J. Kropff, D. Bachelet and H.H. van Laar), 51-66, Cab International, Oxon, UK.

Horie, T., Nakagawa, H. and Ohnishi, M., 1995b. Rice production in Japan under current and future climates. In: *Modelling the Impact of Climate Change on Rice in Asia* (ed by R.B. Matthews, M.J. Kropff, D. Bachelet and H.H. van Laar), 143-164, Cab International, Oxon, UK.

Horie, T., Nakagawa, H., Nakano, J., Hamotani, K. and Kim, H.Y., 1995c. Temperature gradient chamber for research on global environment change. II. A system designed for rice in Kyoto, Japan. *Plant, Cell Environ.*, **18**,1064-1069.

Horie, T., Nakano, J., Nakagawa, H., Wada, K., Kim H.Y. and Seo, T., 1991. Effects of elevated CO_2 and high temperature on growth and yield of rice. 1. Development of temperature gradient tunnels. *Jpn. J. Crop Sci.*, **60** (Extra issue. 2), 127- 128.

Horie, T., de Wit, C.T., Goudriaan, J. and Bensink, J., 1979. A formal template for the development of cucumber in its vegetative stage. *Proc. KNAW, Ser. C*, **82**, 433-479.

Horie, T., Yajima, M. and Nakagawa, H., 1992. Yield forecasting. *Agr. Syst.*, **40**, 211-236.

Imai, K., Colman, D.F. and Yanagisawa, T., 1985. Increase of atmospheric partial pressure of carbon dioxide and growth and yield of rice (*Oryza sativa* L.) *Jpn. J. Crop Sci.*, **54**, 413-418.

Imaki, T., Tokunaga, S. and Obara, S., 1987. High temperature-induced spikelet sterility of rice in relation to flowering time. *Jpn. J. Crop Sci.,* **56** (extra issue 1), 209-210 (in Japanese).

Kim, H.Y., Horie, T., Nakagawa, H., Wada, K. and Seo, T., 1992. Effects of elevated CO_2 concentration and high temperature on growth and nitrogen use efficiency of rice. *Proc. of the First Asian Crop Science Conference: Crop Production and Improvement Technology in Asia.* Korean Society of Crop Science, 205-212.

Matsui, T. and Horie, T., 1992. Effects of elevated CO_2 and high temperature on growth and yield of rice. 2. Sensitive period and pollen germination rate in high temperature sterility of rice spikelets at flowering. *Jpn. J. Crop Sci.,* **61**(extra issue 1), 148-149.

Matthews, R.B., Horie, T., Kropff, M.J., Bachelet, D., Centeno, H.G., Shin, J.C., Mohandass, S., Singh, S., Zhu Defeng and Moon Hee Lee, 1995. A regional evaluation of the effect of future climate change on rice production in Asia. In: *Modelling the Impact of Climate Change on Rice in Asia* (ed by R.B. Matthews, M.J. Kropff, D. Bachelet and H.H. van Laar.), 95-139, Cab International, Oxon, UK.

Monteith, J.L., 1977. Climate and efficiency of crop production in Britain. *Philos. T. Roy. Soc. Lon. B,* **281**, 277-294.

Munakata, K., 1976. Effects of temperature and light on the reproductive growth and ripening of rice. *Proc. of the Symp. on Climate and Rice.,* 187-210, IRRI, Philippines.

Ozawa, Y., 1962. Climatic zoning of Japan for land use. *Chiri,* **7**, Kokinshoin, Tokyo, 300-364 and 432-438 (in Japanese).

Satake, T. and Yoshida, S., 1978. High temperature-induced sterility in Indica rice at flowering. *Jpn. J. Crop Sci.,* **47**, 6-17.

Uchijima, T., 1976: Some aspects of relation between low air temperature and sterile spikelets in rice plants. *J. Agr. Meteorol.,* **31**, 199-202.

Yoshino, M., Horie, T., Seino, H., Tsujii, H., Uchijima, T. and Uchijima, Z., 1988. The effects of climatic variations on agriculture in Japan. In: *The Impact of Climatic Variations on Agriculture, Vol. 1. Assessments in Cool Temperature and Cold Regions* (ed by M.L. Parry, T.R. Carter and N.T. Konijn), 725-865, Kluwer Academic Publishers, Dordrecht.

PREDICTION OF JAPANESE POTENTIAL VEGETATION DISTRIBUTION IN RESPONSE TO CLIMATIC CHANGE

Atsushi Tsunekawa[1], Hitoshi Ikeguchi[2] and Kenji Omasa[1]

ABSTRACT: Potential shift of natural vegetation in Japan caused by global climatic change was predicted by the direct transfer function approach. Two types of vegetation distribution model, fuzzy model and multinomial logit model, were applied to explain the relationship between vegetation classification of remaining natural vegetation and climatic conditions. The logit model indicated a more successful result than the fuzzy model. Then, the effects of increase in mean annual temperature were predicted using the logit model. It was indicated that the percentage of grid-cells in all of Japan which showed different predicted vegetation classifications from the present ones are approximately 23% for a 1-degree increase, 44% for a 2-degree increase, and 62% for a 3-degree increase.
KEY WORDS: global climatic change, natural vegetation, potential shift, direct transfer function approach, vegetation distribution model

1. INTRODUCTION

How does global climatic change affect the vegetation distribution in Japan? Our goal is to answer this question. For this purpose, we have developed statistical models for prediction of the potential vegetation distribution by the direct transfer function approach. Two types of vegetation distribution model, fuzzy model and multinomial logit model, were examined.

What is global climatic change?
The amounts of some trace gases in the atmosphere, especially carbon dioxide (CO_2), nitrous oxide (N_2O), methane (CH_4), chlorofluorocarbons (CFCs) and tropospheric ozone (O_3), have been increasing due to human activities such as the massive consumption of fossil fuels, exhaust emissions from factories and automobiles, agricultural activities, and so on (Japan Environment Agency, 1993). All of these gases are transparent to incoming short-wave radiation from the sun, but they absorb and emit long-wave radiation and are thus able to influence the Earth's climate. This is

---

[1] National Institute for Environmental Studies, Tsukuba, Ibaraki 305, Japan
[2] Museum of Nature and Human Activities, Sanda, Hyogo, 669-13, Japan
This article was expanded from "Effects of climatic change on the vegetation distribution in Japan" by Tsunekawa, A., Ikeguchi, H., Kainuma, M., Tobe, K. and Omasa, K., in *J. Agr. Meteorol.*, 48(5), 867-870, 1993.

called 'global climatic change' or 'global warming' and trace gases which cause this 'greenhouse effect' are called 'greenhouse gases.'

Increased concentrations of CO_2 and other greenhouse gases lead to a warming of the Earth's surface and the lower atmosphere. As a result of the increasing concentrations of greenhouse gases, it is now believed that in the first half of the next century a rise in global mean temperature could occur which will exceed any in man's history (Bolin *et al.*, 1986).

According to the assessment by Working Group I of the Intergovernmental Panel on Climate Change, as a result of the "business as usual" scenario, the average global temperature might have increased by approximately one degree by the year 2025 and by three degrees by the end of the next century. The sea level might rise approximately 20 cm by the year 2030 and 65 cm by the end of the next century (IPCC, 1990a).

Why should we assess the impact on vegetation?

'Global warming' would not only be greater than recent natural fluctuations, but it would occur 15 to 40 times faster than past natural changes (Schneider, 1989). It has been stated by many scientists that such a rate of change may exceed the ability of many species to adapt or disperse into new regions, and thus many plant and animal species may become extinct (IPCC, 1990b). Thus, researches on the impact assessment of climatic change on ecosystems are being promoted all over the world (Goudriaan *et al.*, 1990; Pan, 1993; Peters and Lovejoy, 1992; Rotmans, 1990; Solomon and Shugart, 1993; UK Climate Change Impacts Review Group, 1991).

Dale and Rauscher (1994) summarized the major information needed for forest management.

1. Predictions of expected changes in land use.
2. Compilations of maps of forest type distribution.
3. Compilations of maps of species composition changes.
4. Estimates of changes in forest resource productivity.
5. Estimates of expected changes in forest health.
6. Estimates of potential effect on biological diversity.
7. Predictions of effects on wildlife habitat.

Our target corresponds to the second item; 'Compilations of maps of forest type distribution. Where and when will forest type distributions be altered in response to climate change?'. Strictly speaking, our target may be 'compilations of vegetation type maps of potential distribution in response to climate change.' This is expected to be of critical importance for policy makers and land managers.

Previous studies on Japanese vegetation

Generally speaking, the most important factors limiting vegetation distribution are thermal and moisture conditions. However, because the Japanese islands have abundant precipitation of approximately 1,500 mm/year, thermal conditions limit vegetation more strongly than moisture (Kira, 1974, Yamanaka, 1979). Consequently, plant ecologists have always maintained that Japanese vegetation shows ordination from north to south horizontally, and from high to low altitudinally, according to thermal conditions (Honda, 1922).

To interpret the distribution of Japanese vegetation zones, which is clearly related to thermal conditions, Kira (1949) devised two effective indices, 'the warmth index' (WI) which indicates the positive influence of warmth and the negative influence of excessive heat, and 'the coldness index' (CI) which indicates the negative influence of low temperature. The formulation of these indices is as follows:

$$WI = \Sigma\ (t-5) \text{ for months } t > 5\ ^{\circ}C$$
$$CI = \Sigma\ (5-t) \text{ for months } t < 5\ ^{\circ}C$$

where t is the monthly mean temperature.

Using the warmth index, Uchijima *et al.* (1992) examined which tree species in Japan would be most sensitive to a probable climatic warming.

Why adopt the direct transfer function approach?

The IPCC Working Group II classified the methodologies used for impact studies on ecosystems into four approaches; (1) the direct transfer function approach, (2) the palaeoreconstruction approach, (3) simulation modeling of ecosystems at the stand level, and (4) reasoning from ecological or ecophysiological knowledge (IPCC, 1990b).

We take the first approach; the direct transfer function approach, which has been used by many scientists in impact assessment studies, for example, Emanuel *et al.* (1985a, 1985b) on a global scale, and Zabinski and Davis (1988) for the Great Lakes region of the United States of America.

This method analyzes the current distributions of vegetation types and its bioclimatology (Box, 1981; Desjardins *et al.*, 1992) and predicts their possible future distributions as climatic patterns using the suggested climate change scenarios. As a result of the analysis, it can show areas where current vegetation might be replaced by another type of vegetation in response to climatic change. In these areas it can be considered that the climate is no longer optimal for the current vegetation type. These areas, therefore, are those which would be most vulnerable to disruption as a result of projected climatic changes. As such, considerable care should be given to development of management strategies for these areas (IPCC, 1990b).

2. MATERIALS AND METHODS

2.1 Models

2.1.1 Fuzzy model

The fuzzy model expresses the relationship between vegetation classification and climatic conditions by fuzzy rules with certainty measures. Each rule for estimation of vegetation classification can be described as:

Rule r: if Xr then Ari with Cr,

where Xr is a fuzzy set of climatic conditions such as annual mean temperature, etc., Ari is a subset of vegetation classification, and Cr is a certainty measure of each rule.

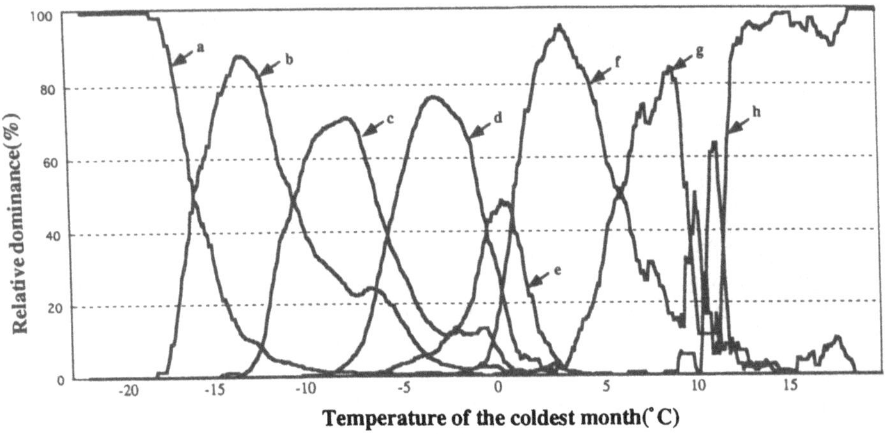

Fig.1. Temperature of the coldest month and relative dominance of each vegetation classification.

Table 1 Examples of fuzzy rules.

If 'Temperature of the coldest month is extremely low' then 'Vegetation classification is a ' with '1.0'.
If 'Temperature of the coldest month is very low' then 'Vegetation classification is b ' with '0.8'
If 'Temperature of the coldest month is low' then 'Vegetation classification is c ' with '0.8'
If 'Warmth index extremely low' then 'Vegetation classification is a ' with '0.8'
If 'Warmth index is very low' then 'Vegetation classification is b ' with '0.8'

The combination of each rule is computed using the probability-theory of Dempster and Shafer (Kainuma *et al.*, 1991).

Figure 1 shows the relative dominance of each vegetation classification for the temperature of the coldest month. *Cr* was determined using such relative dominances. Fuzzy rules could be described as shown in Table 1. Figure 2 shows the structure of rules to estimate vegetation classification.

2.1.2 Linear multinomial logit model

The multinomial logit model is one of several random utility models, which assumes that the doer selects an alternative that has the highest utility and that the utility itself is not constant, but random (Sugiura, 1989). Generally speaking, the relationship between a land and the vegetation on it includes some uncertainty, since the formation process of a plant community involves uncertain elements, for example, competition at the individual level, seed dispersal etc. and also there are some environmental

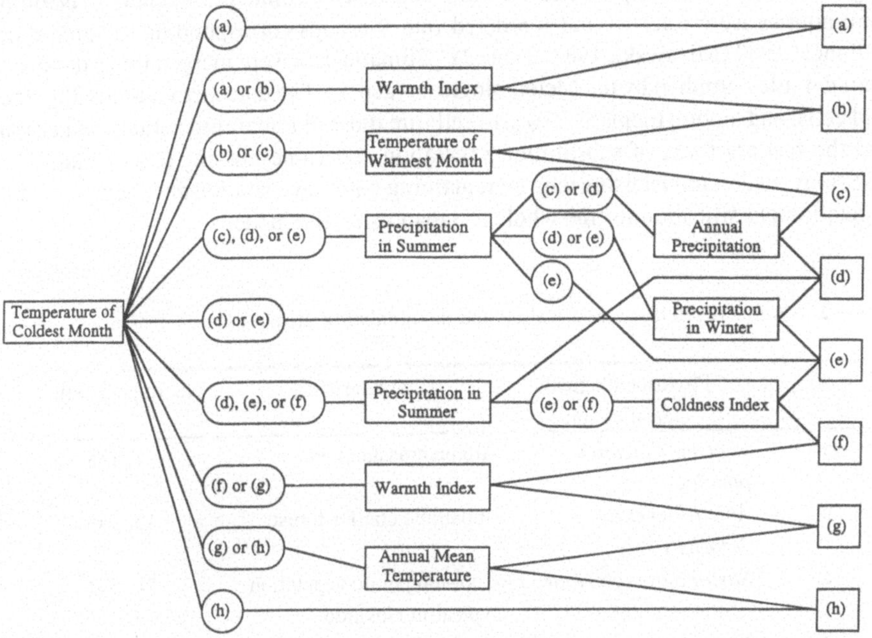

Fig.2. Structure of fuzzy rules for vegetation classification.

conditions which are too difficult to measure or too complex to be included in the analysis. Therefore the random utility model, which assumes the existence of an unknown part for the analysts, would be applicable.

In the case of the linear multinomial logit model, the probability of selecting an alternative is explained to be in proportion to the exponential of the linear combination of the explanatory variables (Ikeguchi *et al.*, 1991).

The multinomial logit model can be shown as

$$P(i)=\left\{\exp\left(\sum_{j=1}^{J}\left(A_{ij}\times V_j\right)+C_i\right)\right\}\bigg/\left\{\sum_{i=1}^{I}\exp\left(\sum_{j=1}^{J}\left(A_{ij}\times V_i\right)+C_i\right)\right\} \tag{1}$$

where $P(i)$: probability of the i-th vegetation classification, V_j: the j-th explanatory variable, A_{ij}: parameter of the j-th explanatory variable of the i-th vegetation classification, C_i: constant of the i-th vegetation.

The study computed the selection probability under each condition from data on vegetation classification and climatic conditions, and then computed the parameters of linear function by the maximum likelihood method.

2.2 Data

Data are arranged in the form of a 1km^2 grid system covering the whole of Japan.

Vegetation data were provided by the Japan Environment Agency. Natural communities were selected and reordered into 8 groups corresponding to "order" or "alliance" level (Miyawaki, 1983, Table 2). Climatic data were merged into a database from datafiles supplied by the Meteorological Agency. The database contains 386,950 grid cells, and among them, 59,786 grid cells are those of remaining natural vegetation and the rest are those of substitution vegetation and value-missing (Color Plate 16). The study made a model using data on remaining natural vegetation and then predicted vegetation classification for the whole of Japan using the model.

Table 2 Natural vegetation classification and the number of grid-cells.

Code	Phytosociological classification	Characteristics	No. of grid-cells
a	*Vaccino-Pinetalia pumilae*	Alpine grassland, etc.	1, 158
b	*Abieti-Piceetalia Jezoensis*	Subalpine conifer forest	15, 345
c	*Saso-Fagion crenatae* (1)	Cool-temperate vegetation typical in Hokaido	24, 541
d	*Saso-Fagion crenatae* (2)	Cool-temperate vegetation typical in the Japan Sea side	13, 542
e	*Sasomoropho-Fagion crenatae*	Cool-temperate vegetation typical in the Pacific side	1, 920
f	*Illicio-Quercetalia acutae*	Warm-temperate vegetation typical on hills	1, 726
g	*Maeso japonicae-Castanopsion Sieboldii*	Warm-temperate vegetation typical on lowlands	657
h	*Psychotrio-Castanopsion Sieboldii*	Subtropical vegetation	897

3. RESULTS

3.1 Estimation by fuzzy model

The hit ratio of the fuzzy model was approx. 70.3% (42,002/59,786). The hit ratios of each vegetation classification were a) 17.7%, b) 49.7%, c) 82.5%, d) 78.2%, e) 39.4%, f) 73.5%, g) 65.9%, h) 99.4%, and their average was 63.2%. The hit ratios of h) and c) were relatively high and those of a) and e) was relatively low. Comparison of the estimated natural vegetation map by the fuzzy model (Color Plate 17) and the actual vegetation map (Color Plate 16) showed that the boundary between cool-temperate vegetation and warm-temperate vegetation of the estimated map was located southward to that of the actual map.

3.2 Estimation by logit model

The hit ratio of the logit model was approx. 75.8% (45,320 /59,786). The hit ratios of each vegetation classification were a) 28.8%, b) 61.7%, c) 78.6%, d) 91.4%, e) 77.2%, f) 83.1%, g) 69.6%, h) 54.4%, and their average was 68.1%. The hit ratios of d) and f) were relatively high and those of a) and h) were relatively low. Comparison of the estimated natural vegetation map by the logit model (Color Plate 18) and the actual vegetation map showed that cool-temperate vegetation of the Pacific side was underestimated.

3.3 Comparison of fuzzy model and logit model

Comparison of the fuzzy model and the logit model showed that both total hit ratio and average hit ratio of the logit model were higher than those of the fuzzy model. From the viewpoint of geographical distribution, those of the logit model were superior to those of the fuzzy model, especially for the discrimination of the Pacific side vegetation and the Japan Sea side vegetation for cool-temperate vegetation.

3.4 Prediction of potential shift using the logit model

By varying the climatic parameters of the logit model, potential shift under the warmed climate was predicted. Three scenarios of 1-degree, 2-degree, and 3-degree increases in annual mean temperature were used. As a result, it was indicated that the percentage of grid-cells in all of Japan which showed different predicted vegetation classifications from the present classifications were approx. 23% for a 1 degree increase, 44% for a 2 degree increase, and 62% for a 3 degree increase. Moreover, the alpine and subalpine vegetation which is distributed separately in the mountainous regions at the present time may be deprived of its suitable habitat (Color Plate19-21).

4. DISCUSSION

From the results of the analysis, it is indicated that even only a 1 degree increase in annual mean temperature might cause potential shift to another vegetation classification for approx. 23% of the total grid-cells (approximately 1km x 1km) in Japan.

What does the potential shift indicated by the study mean? First, as for the accuracy, since the hit ratio of the model is approx. 76%, it must be noted that a corresponding error may be included. Next, as for the ecological meaning of the potential shift, as many ecologists have pointed out, the shifting speed of plant species may be slower than that of a habitat by global warming. Besides, it differs depending on each species. Therefore, it cannot be considered that all plant communities shift to habitats with the same species composition and the same forest structure.

In the regions where shifting of plant species cannot catch up with that of the habitat, some change might be caused in the ecosystem. However, the methodology used in the study has limitations in predicting whether the change might be "competitive displacement" or "catastrophic decline" (Neilson and King, 1991).

Moreover, the model doesn't include all environmental changes in the future. Especially, fertilization effects caused directly by an increase of atmospheric CO_2

might change plant response to temperature, precipitation, and other climatic conditions.

REFERENCES

Bolin, B., Döös, B.R., Jäger J. and Warrick, R.A. (eds.), 1986. *The Greenhouse Effect Climatic Change and Ecosystems.* John Wiley & Sons, Chichester, 541 pp.

Box, E.O., 1981. *Macroclimate and Plant Forms: An Introduction to Predictive Modeling in Phytogeography.* Dr W. Junk Publishers, The Hague, Netherlands, 258 pp.

Dale, V.H. and Rauscher, H.M., 1994. Assessing impacts of climate change on forests: the state of biological modeling. *Clim. Change,* **28**, 65-90.

Desjardins, R.L., Gifford, R.M., Nilson, T. and Greenwood, E.A.N., 1992. *Advances in Bioclimatology.* Springer-Verlag, Berlin, 157 pp.

Emanuel, W.R., Shugart, H.H. and Stevenson, M.P., 1985a. Climatic change and the broad-scale distribution of terrestrial ecosystem complexes. *Clim. Change,* **7**, 29-43.

Emanuel, W.R., Shugart, H.H. and Stevenson, M.P., 1985b. Response to comment: climatic change and the broad-scale distribution of terrestrial ecosystem complexes. *Clim. Change*, **7**, 457-460.

Goudriaan, J., Keulen, H. van. and Laar, H.H. van (eds), 1990. *TheGreenhouse Effect and Primary Productivity in European Aagro-ecosystems.* Pudoc. Wageningen, 90 pp.

Honda, S., 1922. *Studies on Forest Zones in Japan.* Tokyo (in Japanese).

Ikeguchi, H., Takeuchi, K. and Tsunekawa, A., 1991. Potential natural vegetation mapping of Japan by means of GIS: *Environmental Change and GIS. Vol.1* (ed. by I. Ota), 26-33.

Intergovernmental Panel on Climate Change (IPCC), 1990a. *Climate Change, the IPCC Scientific Assessment.* (ed by J.T. Houghton, G.J. Jenkins, and J.J. Ephraums) University Press, Cambridge, 365 pp.

Intergovernmental Panel on Climate Change (IPCC), 1990b. *Climate Change, the IPCC Impacts Assessment.* (ed by W.J. McG. Tegart, G.W. Sheldon and D.C. Griffiths), Australian Government Publishing Service, Canberra, 278pp.

Japan Environment Agency (Global Environment Department), 1993. *Dictionary for Global Environment.* Chuou-Houki, Tokyo, 175 pp. (in Japanese),.

Kainuma, M., Nakamori, Y., Morita, T., Kai, K. and Tsunekawa, A., 1991. Estimation of Japanese vegetation zones using fuzzy modeling. *Proc. 7th Fuzzy System Symposium* (Nagoya, Jun. 12-14, 1991), 561-562 (in Japanese with English abstract).

Kira, T., 1949. *Forest Zones in Japan. Ringyo Kaisetsu Series, 17,* Nippon Ringyo Gijutsu Kyokai, Tokyo (in Japanese).

Kira, T., 1974. A climatological interpretation of Japanese vegetation zones. In: *Vegetation science and environmental protection,* (ed by A. Miyawaki and R. Tuexen), 21-30, Maruzen, Tokyo.

Miyawaki, A.(ed.), 1983. *Handbook of Japanese vegetation* (revised edition), Shibundo, Tokyo, 872 pp. (in Japanese).

Neilson, R.P. and King, G.A., 1991, *Continental Scale Biome Responses to Climatic Change.* The EPA Environmental Research Laboratory in Corvallis, Oregon, 31pp.

Pan, T., 1993. The climate change space influences agricultural production in northeast China. *J. Agr. Meteorol.* **48**, 767-770.

Peters, R.L., Lovejoy, T.E. (eds),1992. *Global Warming and Biological Diversity.* Yale University Press, New Haven & London, 386 pp.

Rotmans, J., 1990. *IMAGE - An Integrated Model to Asses the Greenhouse Effect.* Kluwer Academic Publishers, Dordrecht, 289 pp.

Schneider, S.H., 1989. The greenhouse effect, science and policy. *Science,* 243, 771-781.

Solomon, A.M. and Shugart, H.H. (eds), 1993: *Vegetation dynamics &Global Change.* Champman & Hall, New York, 338 pp.

Sugiura, Y., 1989. *Location and spatial behavior.* Kokon Shoin, Tokyo, 207 pp. (in Japanese).

Tsunekawa, A., Ikeguchi, H., Kainuma, M., Tobe, K. and Omasa, K., 1993. Effects of climatic change on the vegetation distribution in Japan. *J. Agr. Meteorol.,* **48**, 867-870.

Uchijima, Z., Seino, H. and Nogami, M., 1992. Probable shifts of natural vegetation in Japan due to CO_2-climatic warming. *NIAES Series No. 1 (Ecological Processes in Agro-ecosystems),* Natl. Inst. Agro-Environ. Sci., 189-201.

United Kingdom Climate Change Impacts Review Group, 1991. *The Potential Effects of Climate Change in the United Kingdom.* HMSO, London, 124 pp.

Yamanaka, T., 1979. *Japanese forest vegetation.* Tsukiji-shokan, Tokyo, 223 pp. (in Japanese).

Zabinski, C. and Davis, M. B., 1988. Hard times ahead for Great Lakes forests - A climate threshold model predicts responses to CO2-induced climate change -. In: *The Potential Effects of Global Climate Change on the United States,* (ed. by US EPA), 2, 5-1~5-19.

CLIMATIC CHANGE AND ITS IMPACTS ON THE VEGETATION DISTRIBUTION IN CHINA

Atsushi Tsunekawa[1], Xinshi Zhang[2], Guangsheng Zhou[2] and Kenji Omasa[1]

ABSTRACT: The potential vegetation distribution shift in China caused by global climatic change was predicted by the direct transfer function approach. Two types of statistical model, the discriminant analysis model and the multinomial logit model, were applied to explain the relationship between vegetation classification and climatic conditions. The logit model resulted in a more successful result than the discriminant analysis model. Thus, the effects of global climatic change were estimated using the logit model under two climatic change scenarios proposed by Robock *et al.* (1993): Scenario A, a 2-degree increase in annual mean temperature and a 20% increase in annual precipitation; Scenario B, a 4-degree increase in annual mean temperature and a 20% increase in annual precipitation. It was predicted that (1) the conifer forest distributed in northeast China at the present time may be deprived of its habitat within the boundary of China, (2) the broadleaved forest distributed in east China may shift northward by around 3 degrees of latitude for Scenario A and 5 degrees for Scenario B, and (3) the desert region in west China may expand and steppe and savanna may decrease, especially in the case of Scenario B.
KEY WORDS: potential vegetation distribution, climate change, statistical model, impact assessment, China.

1. INTRODUCTION

It is necessary to carry out studies from a variety of approaches to examine the effects of global climatic change on ecosystems. One such approach is to model the relationship between current climatic conditions and actual vegetation and use it to predict potential vegetation distribution resulting from a change in climatic parameters. This kind of approach is called the 'direct transfer function approach', and in the previous section we used it to predict the Japanese potential vegetation. In this section, we attempt the same approach, this time focusing on Chinese vegetation. Here, two types of model, a discriminant analysis model and a linear multinomial logit model, are examined.

---

[1] National Institute for Environmental Studies, Tsukuba, Ibaraki 305, Japan.
[2] Institute of Botany, Chinese Academy of Science, 141 Xizhimenwai Avenue, Beijing, P.R. China.
This article was expanded from "Predicting responses to climatic change of the potential vegetation distribution in China" by Tsunekawa, A., Zhang, X., Zhou, G. and Omasa, K., in *J. Agr. Meteorol.*, 51,245-257,1995.

Why focus on China?

There are some spatial scales for understanding the impact of climate change on vegetation. The implications of the impacts and methodologies for assessing them differ according to the spatial scale to be considered (Frederic, 1994).

Graham *et al.* (1990) divided spatial scales into four categories; biosphere, biome, ecosystem and tree level. The biosphere's responses to climate change involve alterations in the energy, carbon, or water fluxes of vegetation (Henderson-Sellers, 1993; Solomon and Shugart, 1993; MacDonald and Sertorio, 1989). At the biome level, species respond to climate change through evolution, migration, extinction, or adaptation to new disturbance regimes. The ecosystem's responses to climate change take years to centuries and occur via nutrient cycling, production, water use, succession, competition, and response to changes in disturbance regimes (Parton *et al*, 1994). The response of individual trees to climate change occurs through phenological, reproductive, and physiological processes on time scales ranging from minutes to decades and spatial scales ranging from cells to that of a large tree.

The spatial scale of our study corresponds to the biome level in Graham's classification. Its response to climatic changes take years to millennia, and the human activities on this scale relate to plant breeding, land management and conservation.

We have chosen China because it is the largest country in Asia, with an area of approximately 9.6 million km^2. It consists of a wide variety of climatic zones; tropical, sub-tropical, warm-temperate, cool-temperate, alpine, sub-alpine, arid and semi-arid, and the vegetation is also diversified according to environmental conditions (Editorial Committee of Vegetation of China, 1980). Furthermore, China has a human population of approximately 1.2 billion, the largest in the world, and thus its socioeconomic impact is very important.

Previous studies on impact assessment of Chinese vegetation

Ohta *et al.* (1993) simulated changes in the net primary productivity (NPP) of the natural vegetation of monsoon East Asia including China under three, doubled CO_2-climate scenarios (GISS, GFDL and UKMO). The gridded climatic data were used together with the Chikugo model developed by Uchijima and Seino (1985) to assess NPP under baseline and CO_2-doubling climates.

Chang and Yang (1993) analyzed climate-vegetation interaction in China. The distribution and NPP of vegetation zones show a close relationship with a series of climatological indices. Multivariate analysis for climate and vegetation zones in China provides a quantitative environmental interpretation for two significant ecological gradients. The first gradient is mainly a thermal gradient, which can be displayed by latitude, altitude, biotemperature, and annual mean temperature. The second gradient is basically a moisture gradient, and is correlated highly with longitude and potential evapotranspiration.

Chang (1993) related vegetation types and their distribution patterns with certain climatic types in a series of mathematical forms by means of quantitative analysis of vegetation-climate interaction or digitized diagrams of bioclimatology. He applied Holdridge's life zone system in order to relate vegetation and climate and predicted the changing pattern of vegetation under doubled CO_2 conditions.

Our study has some similarity to that of Chang (1993) in that both studies predict impacts caused by climate change with reference to vegetation types in China. Chang (1993) used the Holdridge life zone classification which was originally developed to explain the global vegetation distribution, and the result showed some differences from the actual vegetation distribution. So we attempted to apply statistical prediction models focusing on the Chinese vegetation to improve the accuracy.

This study attempts to answer the question 'How does global climatic change effect the vegetation distribution in China?'. For this purpose, we have developed statistical models for predicting the potential vegetation distribution by the transfer function approach. In Section 2, we examine the climatic trends in China during this century. Then, using the data and methods described in Section 3, the results of model development and prediction are presented in Section 4. An overall discussion follows in Section 5, and finally the conclusions and future steps are described in Section 6.

2. CLIMATIC TRENDS IN CHINA DURING THE 20TH CENTURY

2.1 Data and methods

We analyzed climatic trends in China during the 20th century using 'Global Historical Climatology Network (GHCN) data' compiled by Vose *et al.*, which is a digital database including long-term climatological data on monthly temperature, precipitation, station level air pressure and sea level air pressure (Vose *et al.*, 1992).

We analyzed the data as follows.

1) We extracted temperature and precipitation data for all of China using the country code for China (230). The number of stations was 477 for temperature and 251 for precipitation. Among them, the number of stations for which both temperature and precipitation data were available was 202. To calculate the average values between 1961 and 1990, we searched stations for which both temperature and precipitation data for this period were available. The number of these was 123.

2) By referring to 'Vegetation of China (1980)' etc., we selected 11 representative stations among these 123, and calculated the monthly mean values for temperature and precipitation. The results were shown as a climate-diagram map.

3) Next, we searched stations for which older data were available. We were able to find 17 stations for which both temperature and precipitation data were available since 1911 at the latest, and among these 4 stations had data dating back to 1901. Unfortunately, as shown in Fig. 1, these observation stations were concentrated in eastern China. Therefore, the results of the following analysis should be taken as representative of the climatic trends for eastern China.

We calculated the annual mean temperature and total annual precipitation in each year for each station. However, years in which stations were lacking data, even for one month, were processed as missing years.

4) We subjected the temperature and precipitation data for each station by linear regression analysis and then calculated the fluctuations of temperature and precipitation for each decade.

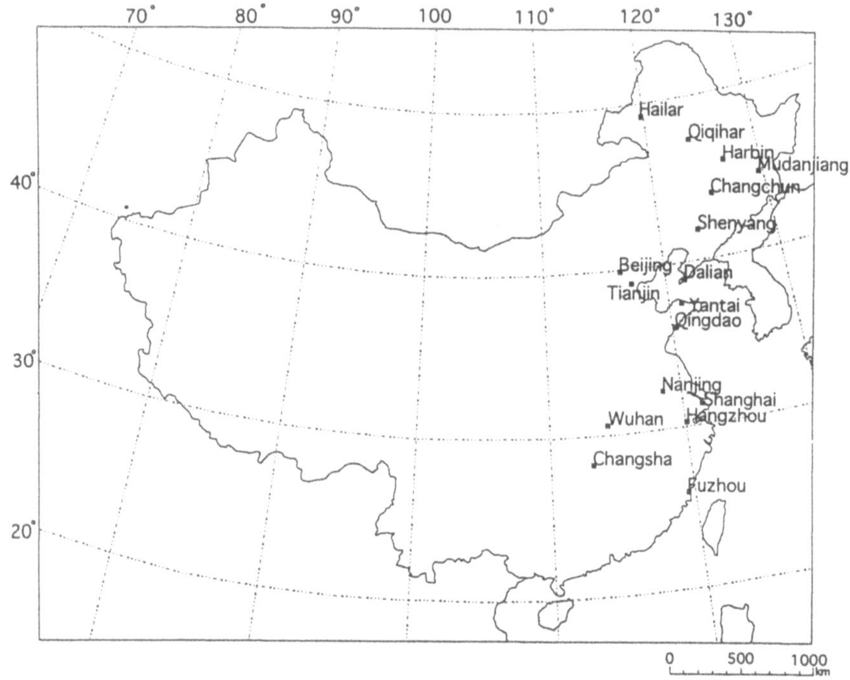

Fig. 1. The distribution of observation stations for which long-term climatological data are available.

2.2 Climate diagram

For understanding the climatic conditions affecting plants at a glance, it is convenient to use a climate diagram developed by Walter *et al.* (Walter and Lieth, 1960-67; Walter *et al.*, 1975). In the diagram, the horizontal axis is changed according to whether the data are for the northern or southern hemisphere, in order to let the center of the figure correspond to the summer. On the vertical axis, the scale of temperature and precipitation is adjusted to let 10°C correspond to 20 mm. The period when the precipitation curve lies below the temperature curve is considered to be a relative dry period.

Among 11 stations representing a variety of Chinese climatic zones, the lowest of annual mean temperature was 2.7 °C at Suihenfe (44.38° N, 131.15° E) and the highest was 21.8 °C at Guangzhou (23.13° N, 113.32° E). The highest annual precipitation was 1700 mm at Guangzhou and the lowest was 235 mm at Urumqi (43.78° N, 87.62° E).

Taking a general view of Fig. 2, we can see a north to south tendency for temperature (north is cold, south is warm) and an east to west tendency for precipitation (east is moist, west is arid).

2.3 Temperature trends

As shown in Table 1, stations which had an inclination of the regression line of more than 0.01, indicating more than a 1°C increase per hundred years and a significance level of more than 5% were Mudanjiang (2.3 °C/century), Shenyang (1.9 °C/c), Hailar

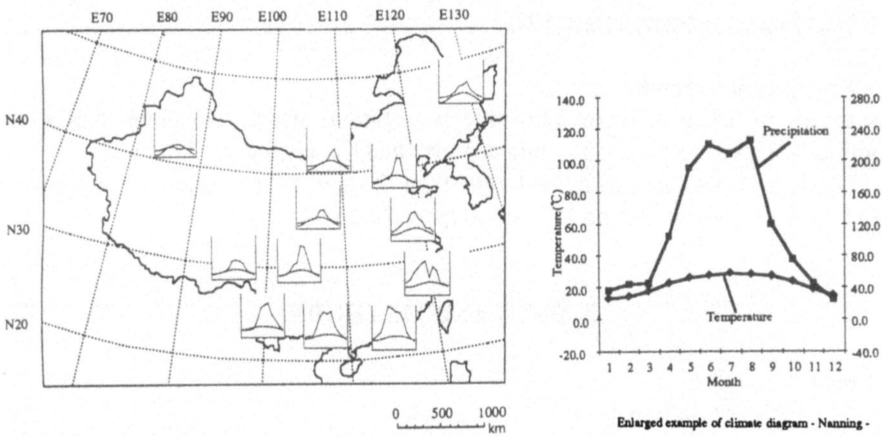

Fig. 2. Climate-diagrams map of China.

Table 1 Climatic trends in the eastern China during the 20th century.

	Latitude (N)	Longitude (E)	Temperature Average (°C)	Temperature Trend (°C/ century)	Temperature Significance level	Precipitation Average (mm)	Precipitation Trend (mm/ century)	Precipitation Significance level
Hailar	49.22	119.75	-2.1	1.4	0.000	315	37	0.318
Qiqihar	47.38	123.92	3.3	0.2	0.608	370	130	0.070
Harbin	45.75	126.77	3.5	1.1	0.000	555	-45	0.396
Mudanjiang	44.57	129.60	3.2	2.3	0.000	516	-3	0.955
Changchun	43.90	125.22	4.8	1.0	0.001	631	-128	0.045
Shenyang	41.77	123.43	7.6	1.9	0.000	685	70	0.333
Beijing	39.93	116.28	11.8	-0.0	0.827	606	18	0.876
Tianjin	39.10	117.17	12.4	-0.0	0.935	527	103	0.154
Dalian	38.90	121.63	10.3	0.4	0.084	592	45	0.598
Yantai	37.53	121.40	12.4	0.6	0.004	661	121	0.088
Qingdao	36.07	120.33	12.1	0.3	0.205	701	171	0.130
Nanjing	32.00	118.80	15.4	-0.0	0.702	1003	70	0.547
Shanghai	31.17	121.43	15.5	0.9	0.000	1158	-29	0.745
Wuhan	30.62	114.13	16.6	-0.8	0.000	1194	88	0.464
Hangzhou	30.23	120.17	16.3	-0.1	0.589	1458	-191	0.118
Changsha	28.23	112.87	17.3	-0.9	0.000	1356	-67	0.613
Fuzhou	26.08	119.28	20.0	-1.5	0.000	1409	-361	0.018

(1.4 °C/c), Harbin (1.1 °C/c) and Changchun (1.0 °C/c).

The only station which had an inclination of the regression line of less than -0.01, indicating more than a 1 °C decrease per hundred years, and a significance level of more than 5% was Fuzhou (-1.5 °C/century).

The increasing tendency was remarkable in northern China, whereas southern China had a decreasing tendency (Table 1).

Figure 3 shows the trends of temperature and precipitation for each decade, in which overall temperature increased during 1911-1940, then dropped around 1941-1950, and again increased from 1951.

2.4 Precipitation trends

As shown in Table 1, there were only two stations which had more than a 5% significance level: Fuzhou (-361 mm/century) and Changchun (-128 mm/c).

Figure 3 shows more complex trends than those of temperature. A drop during 1921-1930 and a peak during 1951-1960 can be seen.

3. DATA AND METHODS

3.1 Models

3.1.1 Discriminant analysis model

Discriminant analysis is a multivariate statistical method, which classifies observations whose group is unknown into one or more groups (nominally measured variables) on the basis of one or more numeric variables (Dillon and Goldstein, 1984).

For the sake of simplicity, if we wish to discriminate a new sample (*e.g.*, each grid cell in the case of this study) into one of two groups (*e.g.*, vegetation classification), G1

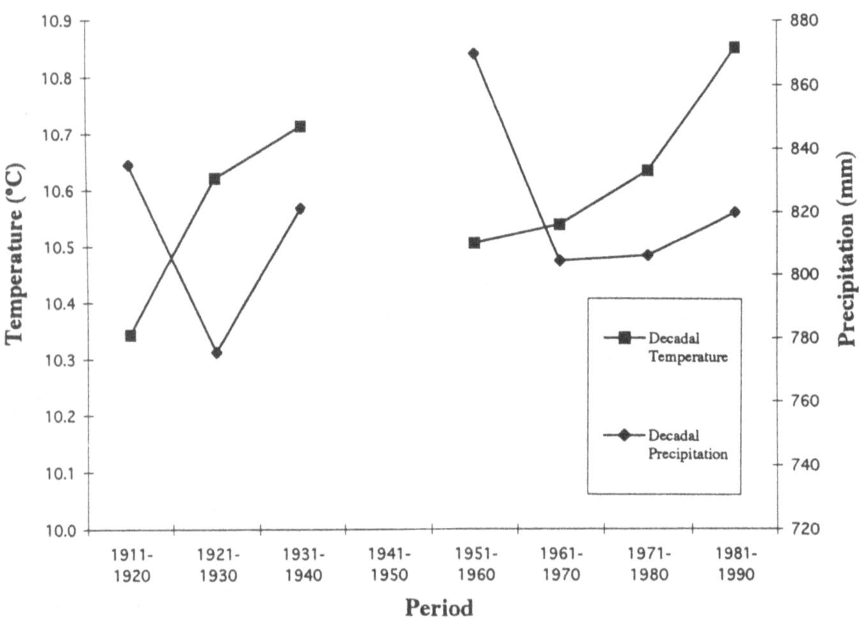

Fig. 3. Decadal fluctuaitons of temperature and precipitation.

and G2, and if we can assume that p independent variables have a multivariate normal distribution and the $p \times p$ variance-covariance matrix of the independent variables in each of the two groups are the same, then letting x_p be the p-th variable (*e.g.*, climatic parameter), σ^{ij} the variance-covariance matrix, and μ_i and μ_j the mean of x_i and x_j, the Mahalanobis distance can be written as

$$D^2 = \sum_{i=1}^{p} \sum_{j=1}^{p} \sigma^{ij}(x_i - \mu_i)(x_j - \mu_j). \tag{1}$$

If the variance-covariance matrix of G1 and G2 is the same and we may define

$$\sigma = (\sigma^{ij}) \text{ and } \bar{\mu} = (\mu_i^{(1)} + \mu_j^{(2)})/2 \tag{2}$$

then

$$D_2^2 - D_1^2 = 2\{a_1(x_1 - \bar{\mu}_i) + a_2(x_2 - \bar{\mu}_2) + \cdots\cdots a_p(x_p - \bar{\mu}_p)\}. \tag{3}$$

The linear discriminant function z is

$$z = a_1(x_1 - \bar{\mu}_1) + a_2(x_2 - \bar{\mu}_2) + \cdots\cdots + a_p(x_p - \bar{\mu}_p) \tag{4}$$

If $z>0$, then we can assign the sample to G1, and if $z < 0$ then we can assign it to G2.

The study used the DISCRIM procedure of "Statistical Analysis System" (SAS release 5.18 at Tokyo University). A likelihood ratio test of the homogeneity of the within-group covariance matrices was performed, and if the test statistic was significant, the within-group matrix was used; otherwise the pooled covariance matrix was used.

3.1.2 Linear multinomial logit model

The multinomial logit model is often used in the field of econometrics to relate environmental conditions to the purchasing behavior of consumers. In the case of the linear multinomial logit model, probability to select an alternative is explained to be in proportion to the exponential of linear combination of the explanatory variables (Ikeguchi *et al.*, 1991). The study computed selection probability under each condition from data on vegetation classification and climatic conditions, and then computed parameters of linear function by the maximum likelihood method.

3.2 Data

Data are arranged in the form of a grid system covering the whole of China in the area of north latitude 18° to 52° and east longitude 73° to 135°. Each grid cell is 150 min for the latitudinal direction and 225 min for the longitudinal direction, corresponding to approximately 5 km x 5 km around central China.

All the data were developed by the Institute of Botany, Academia Sinica. The vegetation data were digitized from the "Actual vegetation map of China (scale 1 : 4,000,000)" made in 1979 by the Institute of Botany. Plant communities were selected and reordered into nine groups corresponding to physiognomy level (Table 2).

We used two climatic variables; annual mean temperature and annual precipitation, which were derived from averages from the period 1951 to 1980 and interpolated from station data (point data) to area data.

The database contains around 370,000 grid cells, among which 331,302 grid cells were used, since the rest had values missing for one or more variables.

3.3 Doubled CO_2-climate scenarios

In studies on climate impact assessment, the output from general circulation models (GCMs) of the climate system is often used. However, the use of GCM output for regional level impact assessment has two serious problems (Robock *et al.*, 1993);

(1) Because current state-of-the-art GCMs are unable to simulate accurately even the current seasonal cycle of climate on a regional basis, the reliability of GCM output is not high.

(2) The spatial resolution of GCMs (typical grid spacing of 500 $\times$ 500 km) is too coarse for regional level studies.

Therefore, Robock *et al.* (1993) have developed a new procedure for combining GCM output with climatic information in order to produce scenarios. They provided three regional scenarios as case studies; China, Sub-Saharan Africa and Venezuela. They showed in the Chinese case study that a warming of 2-4 degrees and an increase in precipitation of about 20% would seem the most likely result of an equivalent doubling of CO_2.

On the basis of their results, we assume two scenarios, Scenario A involving a warming of 2 degrees and a 20% increase in precipitation, and Scenario B involving a warming of 4 degrees and a 20% increase in precipitation.

4. RESULTS

4.1 Estimation by discriminant analysis model

The hit ratio of the discriminant analysis model was approx. 66.1%. The hit ratios of each vegetation classification were, respectively, (a) 77.7%, (b) 54.1%, (c) 64.5%, (d) 55.4%, (e) 79.6%, (f) 66.9%, (g) 69.4%, (h) 45.3% and (i) 67.1%. The hit ratios of (e) and (a) were relatively high and that of (h) was relatively low. Comparison of the estimated natural vegetation map by the discriminant analysis model and the actual vegetation map showed that the vegetation of (b) was underestimated (number of estimated grid cells / actual grid cells = 74%), and the vegetation of (c) overestimated (estimated / actual = 158%).

4.2 Estimation by logit model

The parameters of the logit model are shown in Table 3. For example, the probability that one grid cell with an annual mean temperature of T ℃ and annual precipitation of P mm is assigned to vegetation (a) is;

Table 2 Vegetation classification and their characteristics.

Vegetation code	Physiognomy of vegetation	Dominant plant species	Climatic type by Köppen classification	Distribution	Annual mean temperature (°C)	Annual precipitation (mm)
a	Conifer/Steppe	*Larix* spp., *Pinus* spp., *Arenaria musiformis*	Dw/H	Northeast China (Daxinganling) / Xizang(Tibet)	-2.9 ± 3.4	248 ± 179
b	Meadow/ Broadleaved deciduous forest	*Carex* spp., *Kobresia* spp.	Dw/H	Xizang(Tibet)/ Northeast China (Heilongjiang)	2.1 ± 3.3	510 ± 180
c	Steppe/Desert	*Stipa* spp., *Carex* spp.	BS	Nei Mongol (Chilin, Heilongjiang) /Xinjiang	3.6 ± 2.6	193 ± 106
d	Steppe/Scrub	*Corylus heterophylla*, *Ostryopsis davidiana*, *Vitex negundo* var. *heterophylla*, *Tamarix* spp.	BS	North China (Gansu, Hebei)	5.7 ± 2.7	415 ± 126
e	Desert	*Sympegma regelii*, *Anabasis salsa*	BW	Xinjiang (Takelamagan Desert)	7.4 ± 2.3	77 ± 58
f	Broadleaved deciduous forest/ Three crops for two years or two crops annually	*Quercus* spp., *Acer* spp., *Ulumus* spp.	Cw	North China (Shandong)/ Central China (Anhui, Henan)	11.2 ± 2.7	720 ± 322
g	Evergreen and deciduous scrub/ Two crops containing upland and rice annually	*Rhododendron* spp., *Melastoma* spp., *Pinus* spp.	Cfa	Central China (Hubei)/ Southwest China (Guizhou)	15.8 ± 1.9	1090 ± 227
h	Broadleaved evergreen forest/ Evergreen conifer/ One or double cropping rice	*Cyclobalanopsis* spp., *Castanopsis* spp., *Lauraceae* spp.	Cfa	South China (Fujian)/ Central China (Jiangxi)	17.5 ± 2.7	1486 ± 299
i	Broadleaved evergreen forest/ Double-cropping rice annually	*Vatica astrotricha*, *Dipterocarpus davidiana*	Cfa	South China (Guangdong)/ Southwest China (Yunnan)	19.8 ± 2.0	1588 ± 382

Table 3 Parameters and hit ratios of multinomial logit model.

Vegetation Code	TEMP	PREC	Const.	Hit/Original grid-cells	Hit ratio(%)
a	-0.292×10	-0.145×10^{-1}	0.459×10^{2}	37,463/51,420	72.9
b	-0.263×10	-0.059×10^{-1}	0.429×10^{2}	41,896/60,520	69.2
c	-0.227×10	-0.221×10^{-1}	0.469×10^{2}	16,811/30,598	54.9
d	-0.214×10	-0.133×10^{-1}	0.436×10^{2}	11,798/27,369	43.1
e	-0.174×10	-0.336×10^{-1}	0.456×10^{2}	36,149/42,092	85.9
f	-0.172×10	-0.068×10^{-1}	0.367×10^{2}	25,971/36,999	70.2
g	-0.102×10	-0.042×10^{-1}	0.249×10^{2}	35,164/42,995	81.8
h	-0.079×10	-0.001×10^{-1}	0.152×10^{2}	9,725/20,686	47.0
i	0	0	0	12,570/18,623	67.5
Total				227,547/331,302	68.7

TEMP: Annual mean temperature(℃), PREC:Annual mean precipitation (mm), Const.: constant.

$$P(a) = \left\{\exp\left(T\times(-0.292\times 10)+P\times(-0.145\times 10^{-1})+0.459\times 10^{2}\right)\right\} \Bigg/ \left\{\begin{array}{l}\exp\left(T\times(-0.292\times 10)+P\times(-0.145\times 10^{-1})+0.459\times 10^{2}\right)\\ +\exp\left(T\times(-0.263\times 10)+P\times(-0.059\times 10^{-1})+0.429\times 10^{2}\right)\\ +\cdots\cdots+\exp\left(T\times(0.0)+P\times(0.0)+0.0\right)\end{array}\right\} \quad (5)$$

The estimated vegetation classification can be determined as (x) which has the highest probability among ***P***(a) to ***P***(i).

The hit ratio of the logit model was approx. 68.7% (227,547 /331,302). The hit ratios of each vegetation classification were (a) 72.9%, (b) 69.2%, (c) 54.9%, (d) 43.1%, (e) 85.9%, (f) 70.2%, (g) 81.8%, (h) 47.0% and (i) 67.5%, respectively. The hit ratios of (e) and (g) were relatively high and those of (d) and (h) were relatively low. Comparison of the estimated natural vegetation map by the logit model (Color Plate 23) and the actual vegetation map (Color Plate 22) showed that the boundary between (a) and (b) in the estimated map was greatly simplified in the south-east area of the Tibetan plateau where many valleys run, compared to the actual vegetation map which shows a very complicated distribution resembling a mosaic. Vegetation (d) was underestimated, and especially the distribution around west Tashinganling could not be estimated.

4.3 Comparison of discriminant analysis model and logit model

Comparison of the discriminant analysis model and the logit model showed that the total hit ratio of the latter was higher than that of the former. From the viewpoint of geographical distribution, those of the logit model were superior to those of the discriminant analysis model, especially for the discrimination of (a) and (b) around the south-eastern part of the Tibetan Plateau. Therefore, the logit model was selected for use in the following predictions.

4.4 Prediction of potential vegetation distribution using the logit model under Scenario A

By varying the climatic parameters of the logit model, the potential shift under the changed climate was predicted. First, Scenario A involving a 2-degree increase in annual mean temperature and a 20% increase in annual precipitation was used. As a result, it was indicated that the proportion of grid cells in all of China which showed estimated vegetation classifications different from the present ones was approx. 33%. The percentage of grid cells assigned to classifications other than present one was relatively high in vegetation (h)(68%), (d)(53%) and (a) (51%). The number of grid cells assigned to other classifications is high in vegetation (a)(26,595), (g)(21,310) and (f)(15,258). From the viewpoint of geographical distribution (Color Plate 24), the conifer forest distributed in northeast China at the present time may be deprived of its habitat within the boundary of China. The broadleaf forest distributed in east China (vegetation f, g, h, i) may shift northward by around 3 degrees of latitude.

4.5 Prediction of potential vegetation distribution using the logit model under Scenario B

Next, Scenario B involving a 4-degree increase in annual mean temperature and a 20% increase in annual precipitation was used. As a result, it was indicated that the proportion of grid cells throughout China which showed estimated vegetation classifications different from the present ones was approx. 53%. The percentage of grid-cells assigned to other classifications is relatively high in vegetation (h)(98%), (d)(89%) and (c)(77%). The number of grid cells assigned to other classifications is high in vegetation (a)(35,258), (g)(31,893) and (b)(27,023). From the viewpoint of geographical distribution (Color Plate 25), the broadleaved forest distributed in east China (vegetation f, g, h, i) may shift northward by around 5 degrees of latitude. Moreover the desert region, vegetation (e), may expand its area, and vegetation (c), steppe, may decrease.

5. DISCUSSION

5.1 Climatological trends in China during the 20th century

The tendency for climatic trends shown in Section 2, *i.e.*, a tendency for temperature to be increased in the north and decreased in the south, and an unclear linear tendency for precipitation, is consistent with the results of previous studies (*e.g.*, Chen *et al.*, 1992; Yatagai and Yasunari, 1994).

Our analysis indicates that the temperature inclination during this century has been between -1.5 and 2.3 °C/century at the 5% significance level, depending on the region. The predicted global warming rate is approximately 3 °C/century. Therefore the rate of past warming in China is less than that for predicted global warming, and it should be noted that the region where temperature has been increasing does not cover all of

China.

Therefore, there is a possibility that in the next century when the temperature increase of 3 °C is expected , the ecosystem in China may change more than that in the past century.

5.2 Modeling of vegetation distribution

As a result of the logit model, the total hit ratio is 68.7%. The total hit ratios of Japanese vegetation distribution models developed by the authors were, respectively, 85.4% in the case of the four-group classification corresponding to vegetation zone levels (Takeuchi *et al.*, 1991) and 75.8% for the eight-group classification corresponding to "order" or "alliance" levels (Tsunekawa *et al.*, 1993).

The hit ratio depends strongly on the number of classified groups: the more classified, the lower the hit ratio. Therefore the accuracy of the logit model, with a hit ratio of 68.7% for nine vegetation groups, can be judged as satisfactory.

As shown in the results, there are many discrimination errors between vegetation (a) and (b). If we compare Color Plate 22 and Color Plate 23, it is found that there is a big difference in distribution pattern around east Xizang. This region corresponds to the southwest part of the Tibetan Plateau, which is severely dissected by the Jinsha Jiang River, Lancang Jiang River (Mekong River) and Nu Jiang River (Salween River). Because the vegetation is affected by the elevation and the direction of slope, the vegetation distributions of (a) and (b) are spatially mixed like a mosaic.

Also, the estimated distribution of vegetation (h) is rather simplified in comparison with that of the actual vegetation. In the regions where vegetation (h) is distributed, low mountains, hills, valleys and plains are spatially mixed, thus creating a mosaic-like distribution of vegetation (h).

There are two main reasons why the actual vegetation of these regions could not be estimated well. The first is that the spatial resolution and accuracy of climatic data used by the study are not good enough to reflect the very fine and complex climatic spatial pattern of the severely dissected Plateau. The second is that the vegetation distribution in this region depends not only on temperature and precipitation, but also other environmental factors such as solar radiation, wind, snow and soil properties. However, environmental factors other than temperature and precipitation were not included in the model.

Therefore it is expected that the model could be improved greatly by using more spatially detailed and accurate data and including other environmental factors such as solar radiation, wind, snow and soil properties.

5.3 Prediction of potential vegetation distribution

As described in results, it was predicted under climatic change scenarios that (1) the conifer forest distributed in northeast China at the present time may be deprived of its habitat within the boundary of China, (2) the broadleaved forest distributed in east China (vegetation f, g, h, i) may shift northward by around 3 degrees of latitude for Scenario A and 5 degrees for Scenario B, and (3) the desert region, vegetation (e) may expand its area, and vegetation (c), steppe, may decrease, especially in the case of Scenario B.

Figure 4 illustrates the climatic conditions for each vegetation classification taking the annual mean temperature as the horizontal axis and annual precipitation as the vertical axis. Each rectangular ellipse has its center at the average of temperature and precipitation for each vegetation, and the radiuses represent their standard deviations. Two arrows starting from the center indicate the shifts under Scenarios A and B. Vegetation (a), (b), (c), (d) and (e) have rectangular ellipses and vegetation (f), (g), (h) and (i) have oblong ellipses, indicating that the former vegetation has a relatively small standard deviation for precipitation and the latter a relatively small standard deviation for temperature. This may indicate that the temperature increase affects the latter vegetation more than the former vegetation, and may be the reason why the distribution of broadleaved forest in east China may move more than that of other vegetation.

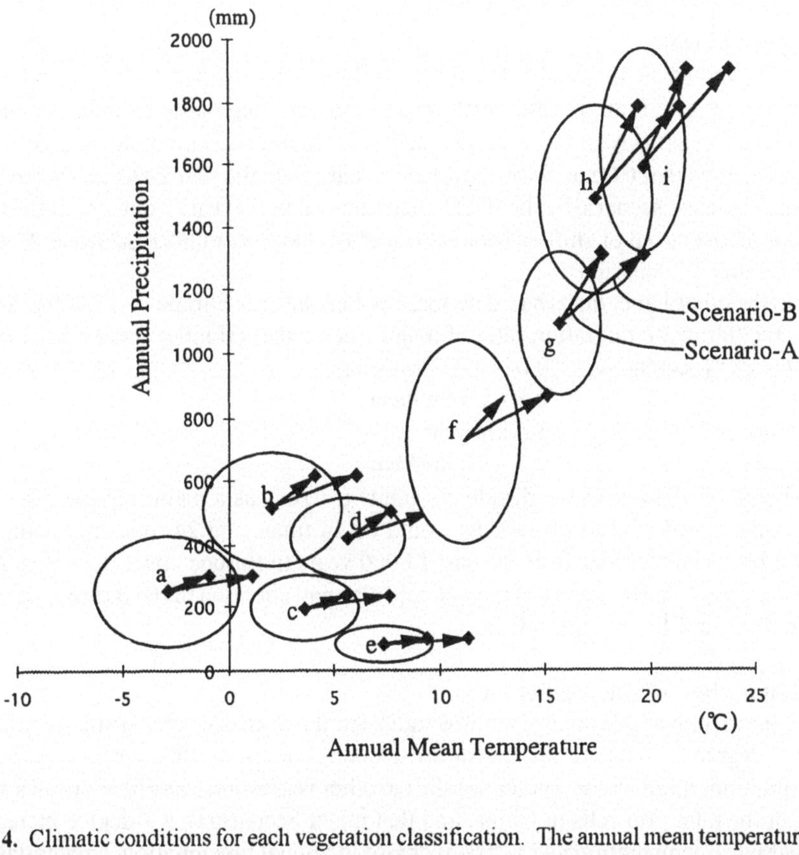

Fig. 4. Climatic conditions for each vegetation classification. The annual mean temperature is taken as the horizontal axis and annual precipitation as the vertical axis. Each rectangular ellipse has its center at the average of temperature and precipitation for each vegetation and the radiuses represent their standard deviations. Two arrows starting from the center indicate the shifts under Scenario A, B.

As shown in Fig. 2, which illustrates the geographical pattern of climate diagrams, southeast China seems to have enough precipitation in summer due to monsoon circulation, whereas west China has a very limited supply of water useful for plants. This fact suggests that thermal condition may be more important for plant growth in the east, whereas moisture conditions may be important in the west.

The most distinctive feature of Scenario B compared to A is that the desert region, vegetation (e), may expand its area, and the area of vegetation (c), steppe, may be reduced. This may be due to the fact that in Scenario B, with a 4-degree increase, even if the precipitation increases by 20%, "drying up" of plants may occur. From the viewpoint of plant physiology, the biggest factors limiting the growth of plants are low temperature and dryness. The increase in temperature is related to both phenomena, but has a positive effect on the former, and a negative one on the latter. It is estimated a 4-degree increase in temperature and a 20% increase in precipitation may have a negative effect on plants through drying up.

5.4 Speed of shift

It was predicted in the study that the broadleaved forest distributed in east China (vegetation f, g, h, i) might shift northward by around 3 degrees of latitude (around 350 km) under Scenario A and by around 5 degrees (around 600 km) under Scenario B. If we can assume that CO_2 doubling will have occurred by the year 2030, as shown in the business as usual scenario by the IPCC, then this will be 50 years from now at the latest. Therefore, the speed of shift of habitats would be 7 km/year under Scenario A and 12 km/year under Scenario B.

On the other hand, maps based on fossil pollen data for Europe 0 - 13,000 years ago show the following migration rates of major tree species (Huntley and Birks, 1983):

Quercus (deciduous)	0.075 - 0.5 km/year
Acer	0.5 - 1.0 km/year
Pinus	1.5 km/year
Ulmus	0.1 - 1.0 km/year

Therefore, if we take the deciduous genus *Quercus* as a main tree species in east China, the speed of shift of habitats would be 14 times and 24 times faster than the migration speed recorded over the past 13,000 years in Europe under Scenarios A and B respectively. At the same time, we should also pay attention to the fact that the speed of shift would differ among species.

5.5 Limitations of the prediction

From the results of this analysis, it was indicated that a climate change under Scenario A, a 2-degree increase in annual mean temperature and a 20% increase in annual precipitation, might cause a potential shift to other vegetation classifications in approx. 33% of the total grid cells in China, and that under Scenario B, a 4-degree increase in annual mean temperature and a 20% increase in annual precipitation, a potential shift to other vegetation classifications for approx. 53% of total grid cells would occur.

What does the potential shift indicated by our study mean?

First, with respect to accuracy, since the hit ratio of the model is approx. 69%, it must be noted that a corresponding error may be included. In particular, the prediction for east Xizang (south-east area of the Tibetan plateau) may not be accurate due to the discriminant errors of the model.

Next, as for the ecological significance of the potential shift, as pointed out above, the actual speed of shift of plant species may be slower than that for a habitat due to climate change. Furthermore, it would differ according to the plant species. Therefore, it appears that all plant communities would not be able to shift to habitats with the same species composition and the same forest structure. In regions where the shift of plant species cannot catch up with that of the habitat, some change would occur in the ecosystem. However, the methodology used in this study has limitations in predicting whether the change might be "competitive displacement" or "catastrophic decline" (Neilson and King, 1991).

Moreover, the model does not include all environmental changes in the future. Especially, fertilization effects caused directly by an increase of atmospheric CO_2 might change the plant responses to temperature, precipitation, and other climatic conditions.

6. CONCLUSIONS AND NEXT STEPS

As we described in the introduction, our final goal is to answer the question 'How does global climatic change affect the vegetation distribution in China?' through compilations of maps of vegetation type distribution. In this study, we were able to identify regions where climatic change might have a serious impact on the actual vegetation. We can summarize the results of the study as follows.

(1) There is a possibility that the ecosystem in China may change in the next century when the temperature increase of 3 °C is estimated more than that in the this century.

(2) As a result of comparison between the discriminant analysis model and the multinomial logit model to explain the actual vegetation distribution in China by annual mean temperature and annual precipitation, the logit model proved superior to the discriminant analysis model in terms of the hit ratios and geographical distribution of estimated actual vegetation.

(3) The total hit ratio for the logit model was 68.7%, which indicates that for nine classes of Chinese vegetation, around 70% of it can be explained by annual mean temperature and annual precipitation. The residuals are thought to be caused by the spatial accuracy of the climatic data used in the study, and the vegetation distribution is considered to depend partially on other environmental factors such as solar radiation, wind, snow, soil properties, and slope gradient.

(4) The potential vegetation under the climatic change was estimated using the logit model. Two scenarios proposed by Robock *et al.* (1993) were applied: Scenario A, a 2-degree increase in temperature and a 20% increase in precipitation, and Scenario B, a 4-degree increase in temperature and a 20% increase in precipitation. As a result, it was predicted under the climatic change scenarios that (1) the conifer forest distributed in northeast China at the present time may be deprived of its habitat within the boundary of China, (2) the broadleaved forest distributed in east China may shift northward by around 3 degrees of latitude for Scenario A and 5 degrees for Scenario B, and (3) the desert region may expand its area, and steppe and savanna may decrease, especially in the case of Scenario B.

If we summarize the impact of climatic change on Chinese vegetation, the habitat for forests might shift rapidly toward the north in east China, and steppe and meadow might be replaced by desert through drying up in west China. Therefore, future research should take into account these results and place priority on the following two points.

(1) In east China, the forest zone might shift in a northerly direction. However, because of the very large population concentration in this region and the predominant use of land in this region for paddy or wheat fields, the majority of forests are isolated geographically. For the forest's smooth shift, flexible dispersal of plant seeds is necessary. However, geographical isolation may prevent the forest from shifting smoothly. Therefore it is urgent priority to develop a 'landscape transition model' based on spatial rules or neighborhood interactions, which can examine the potential for shifting in the face of changing landscape patterns for the purpose of appropriate forest management (Dale and Rauscher, 1994).

(2) In west China, especially the areas surrounding deserts, the change in moisture conditions due to increasing temperature and drying-up may cause 'desertification' of steppe and savanna. Then desertification accompanied by reduction of vegetation may accelerate the climatic change through an increase in albedo, reduction in soil moisture and evapotranspiration, and decrease in surface roughness, *i.e.*, a positive feedback of climatic change. Therefore it is necessary to develop a 'grassland model' that can simulate the regional patterns of ecosystem properties including plant production and soil properties., taking account of fertilization effects by increased CO_2 , extreme events such as drought and severe heat, and livestock management for appropriate grassland management so as not to cause desertification (Conner, 1994). The CENTURY model (Parton *et al.*, 1987) or EPIC model (Williams *et al.*, 1984) may be instructive for development of such a model.

REFERENCES

Chang, H., 1993. A vegetation-climate classification system for global change studies in China. *Q. Sci.*, **2**, 157- 173. (in Chinese with English abstract).

Chang, H. and Yang, D., 1993. Study on climate-vegetation interaction in China. *J. Environ. Sci.*, **5**, 130-149.

Chen, L.-X., Dong, M. and Shao, Y.N., 1992. The characteristics of interannual variations of the East Asian Monsoon. *J. Meteorol. Soc. Jpn.*, **70**, 397-421.

Conner, J.R., 1994. Assessing the socioeconomic impacts of climate change on grazinglands. *Clim. Change*, **28**, 143-157.

Dale, V.H. and Rauscher, H.M., 1994. Assessing impacts of climate change on forests: the state of biological modeling. *Clim. Change*, **28**, 65-90.

Dillon, W.R. and Goldstein, M., 1984. *Multivariate Analysis - Methods and Applications.* John Wiley & Sons, New York, 587 pp.

Editorial Committee of Vegetation of China, 1980. *Vegetation of China.* Science Press, Beijing, 1375 pp. (in Chinese).

Frederick, K.D., 1994. Integrated assessments of the impacts of climate change on natural resources. *Clim. Change*, **28**, 1-14.

Graham, R.L., Turner, M.G. and Dale, V.H., 1990. How increasing atmospheric CO_2 and climate change affect forests. *BioScience*, **40**, 575-587.

Henderson-Sellers, A., 1993. Continental vegetation as a dynamic component of a global climate model: a preliminary assessment. *Clim. Change*, **23**, 337-377.

Huntley, B. and Birks, H.J.B., 1983. *An Atlas of Past and Present Pollen Maps for Europe: 0-13000 Years Ago*. Cambridge University Press, Cambridge, 667pp.

Ikeguchi, H., Takeuchi, K. and Tsunekawa, A., 1991. Potential natural vegetation mapping of Japan by means of GIS. In: *Environmental Change and GIS. Vol.1* (ed. by I. Ota), 26-33.

MacDonald, G. and Sertorio, L. (eds.), 1989. *Global Climate and Ecosystem Change*, Plenum Press, New York, 252 pp.

Neilson, R.P. and King, G.A., 1991. *Continental Scale Biome Responses to Climatic Change*. The EPA Environmental Research Laboratory in Corvallis, Oregon, USA, 31 pp.

Ohta, S., Uchijima, Z. and Oshima, Y., 1993. Probable effects of CO_2-induced climatic changes on net primary productivity of terrestrial vegetation in East Asia. *Ecol. Res.*, **8**, 199-213.

Parton, W.J., Schimen, D.S., Cole, C.V. and Ojima, D.S., 1987. Analysis of factors controlling soil organic levels of grasslands in the Great Plains. *Soil Sci. Soc. Am. J.*, **51**, 1173-1179.

Parton, W.J., Ojima, D.S. and Schimel, D.S., 1994. Environmental change in grasslands: assessment using models. *Clim. Change*, **28**, 111-141.

Robock, A., Turco, R.P., Harwell, M.A., Ackerman, T.P., Andressen, R., Chang, H.-S. and Sivakumar, M.V.K., 1993. Use of general circulation model output in the creation of climate change scenarios for impact analysis. *Clim. Change*, **23**, 293-335.

Solomon, A.M. and Shugart, H.H. (eds.), 1993. *Vegetation Dynamics & Global Change*. Champman & Hall, New York, 338 pp.

Takeuohi, K., Ikeguchi, H. and Tsunekawa, A., 1991. Estimation of Japanese vegetation zones using GIS. *Reports for Scinece Research Budget of the Ministry of Education, Science and Culture, F.Y. 1990, GIS for Environmental Change by Modernization*, 13-20. (in Japanese with English abstract).

Tsunekawa, A., Ikeguchi, H., Kainuma, M., Tobe, K. and Omasa, K., 1993. Effects of climatic change on the vegetation distribution in Japan. *J. Agr. Meteorol.*, **48**, 867-870.

Tsunekawa, A., Zhang, X., Zhou, G. and Omasa, K., 1995. Predicting responses to climatic change of the potential vegetation distribution in China. *J. Agr. Meteorol.*, **51**, 245-257.

Uchijima, Z. and Seino, H., 1985. Agroclimatic evaluation of net primary productivity of natural vegetation. (1) Chikugo model for evaluating net primary productivity. *J. Agr. Meteorol.*, **40**, 343-352.

Vose, R.S., Heim, R., Schmoyer, R.L., Karl, T.R., Steurer, P.M., Eischeid, J.K. and Peterson, T.C., 1992. *The Global Historical Climatology Network - Long-term Monthly Temperature, Precipitation, Sea Level Pressure and Station Level*

Pressure Data. Oak Ridge National Laboratory, Oak Ridge, USA, 311 pp.
Walter, H. and Lieth, H., 1960-67. *Klimadiagramm - Weltatlas* (in 3 Lief.), G. Fischer, Jena.
Walter, H., Harnickell, E. and Mueller-Dombois, D., 1975. *Climate Diagram Maps*, Springer-Verlag, Berlin, 36pp.
Williams, J.R., Jones, C.A. and Dyke, P.T., 1984. A modeling approach to determining the relationship between erosion and soil productivity. *Trans. Am. Soc. Agr. Eng.*, **27**, 129-144.
Yatagai, A. and Yasunari, T., 1994. Trends and decadal-scale fluctuations of surface air temperature and precipitation over China and Mongolia during the recent 40 year period (1951-1990). *J. Meteorol. Soc. Jpn.*, **72**, 937-957.

EFFECTS OF GLOBAL WARMING ON THE PHENOLOGICAL OBSERVATION IN JAPAN

Keiko Kai[1], Mikiko Kainuma[1] and Naomi Murakoshi[2]

ABSTRACT: Strong correlations were found between blooming dates and meteorological factors. On the basis of on these correlations, predictive maps of blooming dates in the Japanese Islands were proposed for each case of 1°C, 2°C and 3°C of monthly mean temperature warming. The correlation was tested for the blooming dates of *Prunus yedoensis, Prunus Mume, Camellia japonica, Taraxacum* sp., *Rhododendron Kaempferi, Wistaria floribunda, Lespedeza bicolor, Hydrangea macrophylla, Lagerstroemia indica, Miscanthus sinensis*, etc., using the data on monthly mean temperatures from 102 meteorological stations in Japan for the period 1953-1990. Simple and multiple regression analyses were used for the correlation.

Among meteorological factors, the strongest correlation was shown for monthly mean temperatures. Notably, the strongest was obtained for the case of *Prunus yedoensis*. The mean temperature of the previous December also showed the best correlation for species such as *Prunus Mume* and *Camellia japonica*. Strong correlations between the leaf color-changing dates of *Ginkgo biloba* and *Acer palmatum* and the monthly mean temperature were found in one month of autumn. In these species, there was a delay of 2-7 days with a 1 degree increase in mean temperature.

The 30-year 1 km^2 temperature-climate mesh-file developed by the Japan Meteorological Agency was used for the phenological estimation and predictive maps of phenological dates. Each observatory station was classified according to its annual mean temperature. Phenological dates for each mesh were estimated through monthly mean temperatures and regression equations of corresponding stations. Then, distribution maps of predictive phenological dates distinguished by 5-day divisions were made.

KEY WORDS: phenology, global warming, mean temperature, blooming date, plant, Japan.

1. INTRODUCTION

Natural biological processes that are good indicators of seasonal change have been themselves changing due to the effects of global warming. According to a recent report by the Japan Meteorological Agency, a significant shift was found in

[1] National Institute for Environmental Studies, Tsukuba, Ibaraki 305, Japan.
[2] Faculty of Science, Shinshu University, Asahi 3-1-1, Matsumoto, Nagano 390, Japan.

phenology between the thirty-year averaged period of 1956-1985 and 1961-1990 of the Manual of Phenological Observation (JMA, 1991). The secular change in blooming and sprouting dates in Japan has a close relation to meteorological indices such as monthly mean temperature, precipitation, solar radiation, monthly maximum temperature, monthly minimum temperature, and humidity, especially with monthly mean temperature (JMA, 1988). The blooming date is controlled by the progression of temperature change before the blooming date (Shigehara, 1991; Yoshino *et al.*, 1995). The blooming date of *P. yedoensis* shifts by 4 days with a 1°C change of temperature in March (Ohshima, 1987). Inversely, an attempt was made to estimate the blooming dates of plants using the method of simple and multiple regression analyses, degree-day and degree-hour, the number of days transformed to standard temperature, latitude, distance from shoreline, cold index, and warmth index (Omoto and Aono, 1989, 1990, 1991; Aono and Omoto, 1990a, b). Since 1987 JMA has been reporting forecasts of blooming dates of *P. yedoensis* by means of simple and multiple regression equations, formula of bud-weight, long-term prediction, analog of year, variables of mean temperature, lowest temperature, and precipitation etc.

We estimated the dates of blooming, budding, leaf-color change, and leaf falling affected by global warming using simple temperature data. This would be a good strategy for the study of indicators of shifting of seasons and environmental change, estimation of phenological phenomena in places where meteorological data are absent, investigation of the climate of historical ages, application for agricultural aspects such as budding, blooming, harvesting, and utility in recreation and sightseeing.

2. MATERIALS AND METHOD

Observation of phenology has been carried since 1953 based on the "Manual of Phenological Observation" that suggests the observatory method and its provisions (JMA, 1988). Using these phenological data, the relation between the blooming dates of each plant (*P. yedoensis*, *P. Mume*, *Taraxacum* sp., *Wistaria floribunda*, etc.) and the monthly means of meteorological factors (mean temperature, precipitation, solar radiation, maximum temperature, minimum temperature, humidity, etc.) were investigated. We tried to estimate the shift of phenology under the warming effect using monthly mean temperature, because the monthly mean temperature shows the significant correlation with the phenological dates of each plant.

Correlations between the phenological dates and monthly mean temperatures at each station (102 localities) were examined and simple and multiple regression equations were calculated as a function of correlated monthly mean temperature. Shifts of blooming day due to temperature rise were obtained for each station from this formula. The values of the shift at each station were plotted on the Japanese map with the advantage of the 30-year 1 km temperature-climate mesh-file. First, each observatory station was classified by a 1°C temperature according to its annual mean temperature. Second, each 1 km^2 mesh-file was integrated into km^2 scaled monthly and annual mean temperatures. By substituting these values into the equations of each station within the 5x5 km grid, blooming dates were estimated. Distribution maps of the blooming dates were made by assuming that the mean values of the estimated phenological dates were those of each grid.

3. RESULTS

3.1 Estimation of the blooming date of *Prunus yedoensis*

On the most effective over 2-3 month periods, correlation of blooming dates and monthly mean temperatures were examined for the 38 years from 1953 to 1990. There were strong correlations for the previous month of blooming day or the month including the blooming day. Regression equations were calculated at each station. The results show that the blooming dates before April 21 have high correlation with the monthly mean temperature of March (mean correlation coefficient of 0.773); the blooming dates after April 20 have high correlation with those of April (mean correlation coefficient of 0.806). According to multiple regression analysis for two months, the mean correlation coefficient is about 0.8 for March and April, 0.9 for April and May.

Overall, a 1°C rise of monthly mean temperature of March leads to 2.7-4.8 days of blooming with a mean of 3.24 days. There were local differences in blooming under the warming effect: In southern Kyushu, southern Shikoku, the Goto and Bonin Islands, correlation coefficients became lower when the monthly mean temperature was above 10°C. If the monthly mean temperature was higher than 12°C there was almost no correlation. The shift when the monthly mean temperature rose by 1°C was about 2.5 days for these regions. On the other hand, the shift was more than 4 days in Hokkaido. Distribution maps shifting the blooming dates of each station in the cases of 1°C, 2°C and 3°C rises in monthly mean temperature were made for the 10-day intervals before March 20, March 21 to 30, March 31 to April 9, April 10 to 19 and April 20 to 29, April 30 to May 9 and after May 10.

Color Plate 26a shows the 30-year (1951-1980) mean blooming dates. The light-pink colored areas, which spread between 31°34'N-36°33'N, indicate the regions where the blooming dates were from March 21 to March 30. The Kanto Plain, the Pacific side of Honshu and some parts of Shikoku and Kyusyu are included within this area. Red colored areas are the places where the blooming dates were before March 20. An estimated shift of blooming date under a 3°C rise in mean temperature in March is shown in Color Plate 26b. Light-pink colored areas shifted northward including Sendai (38°16'), Aikawa, Niigata and Wajima. The pattern of distribution appears very similar to the pattern of March 31 to April 9 for the 30-year mean. The blooming front (Sakura Zensen, which is an analogy of meteorological front moving northwards in spring) shifted about 3° in latitude relative to that of the 30-year means. At the end of the 21st century, it is expected that the global temperature will rise by 3°C (IPCC, 1990). This simulation indicates that Sakura Zensen shifts northward from northern Kanto to the Tohoku and Hokuriku regions, and the blooming dates in the region south of northern Kanto shifted within the March 17 to 26 period.

3.2 Estimation of the blooming date of *Prunus Mume*

The same analysis was carried out for *P. Mume* is classified as a warming-fuddled type by JMA (1991). However, the results indicate that the correlation between the blooming dates of *P. Mume* and the mean temperature of two or three month interval previous to blooming is as high as that of *P. yedoensis*. The values of correlation coefficients were relatively higher at the stations in northern Japan (r=0.85 in north, while r=0.68 is the mean for Shikoku and Kyusyu). The correlation decreases if the monthly mean temperatures are above 12°C or below 5°C. The blooming dates shifted by 4-13 days (about 6.8 days on average) with a 1°C monthly mean temperature rise of January .

Distribution maps of blooming dates of each station were made for 20-year means and for the cases of 1°C, 2°C and 3°C rise in monthly mean temperature of December, January, February, March and April, for the 30-day intervals before January 19, January 20 to February 8, February 9 to 28, March 1 to 20, March 21 to April 9, April 10 to 29, and after April 30 (Color Plate 27). Various subspecies are included in this case. Red and light-pink colored areas indicate the distribution of the blooming dates before February 8. The estimated distribution map in the case of 2°C rise in monthly mean temperature is quite similar to the map for blooming dates from March 28 to April 9 for the 30-year mean.

3.3 Estimation of the blooming dates of *Taraxacum* sp.

Estimation of blooming dates for *Taraxacum* sp. was carried out. *Taraxacum* sp. is classified as a warming-adaptation type by JMA (1991). The correlation between blooming dates of *Taraxacum* sp. and monthly mean temperatures for two or three month interval is not as high as those of *P. yedoensis* and *P. Mume*. However, a similar trend of correlation coefficients with *P. Mume* was found; The highest value (r=0.67) was observed in northern Japan and the value decreased in southern Japan. The correlation decreases if the monthly mean temperature is above 10°C or below 5°C. The blooming dates shift 3-8 days (about 4.5 days on average) earlier with 1°C monthly mean temperature rise of February. Distribution maps of blooming dates for each station were made for the 30-year mean and for the cases of 1°C, 2°C and 3°C rises in monthly mean temperature in December, January, February, March, April and May, for the 30-day intervals from January 30 to February 18, February 19 to March 10, and March 11 to 30, March 31 to April 19 and April 20 to May 9 (Color Plate 28).

In the distribution map for the 30-year mean, the blooming dates were within March except for the mountainous region in southern Kanto. In the estimated distribution map, 3°C rise causes that the blooming dates at the stations south of Tohoku shifts from March to February.

3.4 Estimation of the blooming date of *Camellia japonica*

The relation of mean temperature with blooming date of *Camellia japonica* L. was examined. The results indicate that the correlation between the blooming date of *C. japonica* and the mean temperature of two or three months interval is not as high as that of *P. yedoensis,* however, similar trend was also found. The highest correlation (r=0.91) was observed in northern Japan and the values decreased in southern Japan. The blooming dates shift 5-15 days (about 7.2 days on average) earlier with 1°C monthly mean temperature rise of January. The blooming dates shift 5-18 days (about 9.3 days on average) earlier with 1°C monthly mean temperature rise of December. Distribution maps of blooming dates for each station were made for the 30-year mean and for the case of 1°C, 2°C, and 3°C rise in monthly mean temperature of December, January, February, March, and April, for the 10-days intervals of December 13 to January 1, January 2 to 21, January 22 to February 10, February 11 to March 2, and March 3 to 11 (Color Plate 29). The magnitude of shift of blooming day with 3°C mean temperature warming becomes smaller in the northern limit of the distribution.

3.5 Estimation of the blooming date of *Rhododendron Kaempferi*

The blooming dates shift 3-15 days (about 3.1 days on average) earlier with 1°C monthly mean temperature rise of April. The blooming dates shift 2-6 days (about 3.2 days on average) earlier with 1°C monthly mean temperature rise of March. Distribution maps of blooming dates for each station were made for the 30-year mean and for the case of 1°C, 2°C, and 3°C rise in monthly mean temperature of February, March, April, and May, for the 10-day intervals of April 6 to 15, April 16 to 25, April 26 to May 5, May 6 to 15, and May 16 to 25 (Color Plate 30).

3.6 Estimation of the blooming date of *Wistaria floribunda*

The blooming dates shift 2-5 days (about 3.2 days on average) earlier with 1°C monthly mean temperature rise of April. The blooming dates shift 2-4 days (about 2.4 days on average) earlier with 1°C monthly mean temperature rise of March. Distribution maps of blooming dates for each station were made for the 30-year mean and for the case of 1°C, 2°C, and 3°C rise in monthly mean temperature of February, March, April, and May, for the 10-day intervals of April 10 to 19, April 20 to 29, April 30 to May 9, May 10 to 19, and May 20 to 29 (Color Plate 31).

3.7 Estimation of the blooming date of *Hydrangea macrophylla*

The blooming dates shift 3-7 days (about 4.1 days on average) earlier with 1°C monthly mean temperature rise of May. Distribution maps of blooming dates for each station were made for the 30-year mean and for the case of 1°C, 2°C, and 3°C rise in monthly mean temperature of March, April, May, June, and July, for the 20-day intervals of May 30 to June 18, June 19 to July 8, July 9 to 28, July 29 to August 17, and August 18 to September 6 (Color Plate 32).

3.8 Estimation of the blooming date of *Hydrangea maritima*

Distribution maps of blooming dates for each station were made for the 30-year mean and for the case of 1°C. 2°C, and 3°C rise in monthly mean temperature of

March, April, May, June, and July, for the 20-day intervals of May 11 to 30, May 31 to June 19, and June 20 to July 9 (Color Plate 33).

3.9 Estimation of the blooming date of *Miscanthus sinensis*

Distribution maps of blooming dates for each station were made for the 30-year mean and for the case of 1°C, 2°C, and 3°C rise in monthly mean temperature of July, August, September, and October, for the 10-day intervals of August 15 to 24, August 25 to September 3, September 13, September 14 to 23, and September 24 to October 3 (Color Plate 34).

3.10 Estimation of the budding dates of *Ginkgo biloba*

The blooming dates shift 1-6 days (about 4.0 days on average) earlier with 1°C monthly mean temperature rise of March. Distribution maps of budding dates for each station were made for the 30-year mean and for the case of 1°C, 2°C, and 3°C rise in monthly mean temperature of February, March and April, for the 10-day intervals of April 6 to 15, April 16 to 25. April 26 to May 5, May 6 to 15, and May 16 to 25 (Color Plate 35).

3.11 Estimation of the dates of leaf-color change and leaf-falling of *Ginkgo biloba*

The dates of leaf-color change and leaf-falling shift 2-7 days (about 4.2 days on average) later with 1°C monthly mean temperature rise of October. Distribution maps of the dates of leaf-color change for each station were made for the 30-year mean and for the case of 1°C, 2°C, and 3°C rise in monthly mean temperature of August, September, October, and November, for the 10-day intervals of October 7 to 16, October 17 to 26, October 27 to November 5, November 6 to 15, November 16 to 25 (Color Plate 36).

The dates of leaf-color change and leaf-falling shift 2-7 days (about 4.3 days on average) later with 1°C monthly mean temperature rise of October. Distribution maps of leaf-falling dates for each station were made for the 30-year mean and for the case of 1°C, 2°C, and 3°C rise in monthly mean temperature of September, October, November, and December, for the 10-day intervals of November 6 to 15, November 16 to 25, November 26 to December 5, December 6 to 15, and December 16 to 25 (Color Plate 37).

3.12 Estimation of the dates of leaf-color change and leaf-falling of *Acer palmatum*

The dates of leaf-color change shift 2-12 days (about 3.7 days on average) later with 1°C monthly mean temperature rise of November. Distribution maps of the dates of leaf-color change for each station were made for the 30-year mean and for the case of 1°C, 2°C, and 3°C rise in monthly mean temperature of August, September, October, and November, for the 10-day intervals of October 15 to 24, October 25 to November 3, November 4 to 13, November 14 to 23, and November 24 to December 3 (Color Plate 38).

The dates of leaf-falling shift 2-12 days (about 3.3 days on average) later with 1°C monthly mean temperature rise of November. Distribution maps of leaf-falling dates for each station were made for the 30-year mean and for the case of 1°C, 2°C, and 3°C rise in monthly mean temperature of August, September, October, November, and December, for the 10-day intervals of Nov. 6 to 15, Nov. 16 to 25, Nov. 26 to Dec. 5, Dec. 6 to 15, and Dec. 16 to 25 (Color Plate 39).

4. DISCUSSION

The results indicate that a high correlation was observed during the relatively low-temperature months of March to April, and that the cold indices have high correlation with the blooming date of *P. Mume*, *Camellia japonica* and *Taraxacum* sp. The blooming dates of these three plants are earlier than that of *P. yedoensis*. Figure 1 shows the average change in blooming dates of the three fruit trees under conditions that monthly mean temperature rises by 1°C due to climatic warming. The magnitudes of the shifts are 7.2 days for *C. japonica*, 6.8 days for *P. Mume* in the case of 1°C rise in monthly mean temperature in January; 4.1 days for *Taraxacum* sp. under 1°C rise in monthly mean temperature in March. These magnitudes of shift are larger than that of *P. yedoensis*. For the plants with blooming dates after March, 3.0 days in shift were found for *Rhododendron Kaempferi* and *Wistaria floribunda* under 1°C rise in monthly mean temperature of April. The accuracy of estimation is the same with both plants (Aono and Omoto, 1992). Monthly mean temperatures for three months were effective on blooming

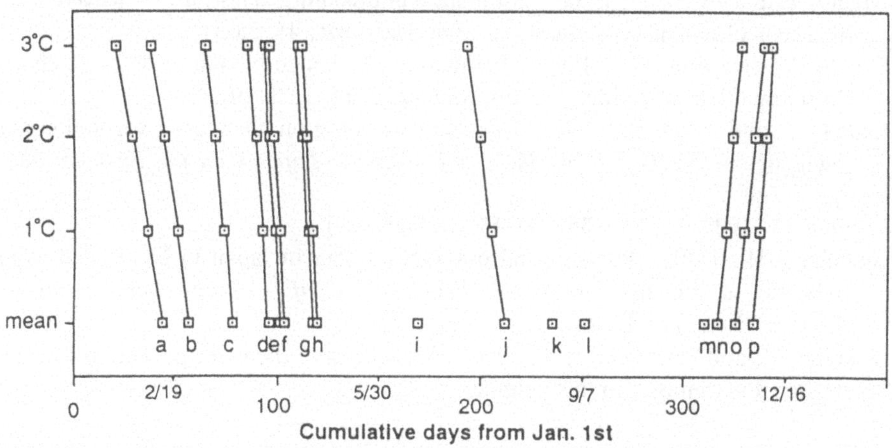

Fig. 1. Shifts in phenological dates for each plant. a: *Camellia japonica*, b: *Prunus Mume*, c: *Taraxacum* sp., d: *Prunus yedoensis* (bloom), e: *Ginkgo biloba* (budding), f: *P. yedoensis* (full bloom), g: *Rhododendron Kaempferi*, h: *Wistaria floribunda*, i: *Hydrangea macrophylla*, j: *Lagerstroemia indica*, k: *Lespedeza bicolor*, l: *Miscanthus sinensis*, m: *G. biloba* (leaf-color change), n: *Acer palmatum* (leaf-color change), o: *G. biloba* (leaf fall), p: *A. palmatum* (leaf fall).

dates of plants from late April; these include *Hydrangea macrophilla*, *Lagerstroemia indica*, and *Miscanthus sinensis*. On *Lagerstroemia indica*, the highest correlated month is June; 1°C rise in temperature makes the blooming days 6.4 days earlier. On *Ginkgo biloba* L. and *Acer palmatum*, high correlation was found in September to November, and the leaf-color change dates and the leaf-falling dates shift about 4 days later under 1°C temperature rise.

REFERENCES

Aono, Y. and Omoto, Y., 1990a. Estimation of blooming date for *Prunus yedoensis* using DTS combined with chill-unit accumulations. *J. Agr. Meteorol.*, **45**, 243-249.

Aono, Y. and Omoto, Y., 1990b. A simplified method for estimation of blooming date for the cherry by means of DTS. *J. Agr. Meteorol.*, **46**, 147-151.

Aono, Y. and Omoto, Y., 1992. Estimation of blooming dates *Phododendron Kaempferi*, *Wisteria floribunda* and *Hydrangea macrophylla*. *J. Agr. Meteorol.*, **47**, 233-240.

IPCC, 1990. *Climate Change - The IPCC Scientific Assessment.* Cambridge University Press, 366pp.

Japan Meteorological Agency, 1988. Phenological observation in Japan. *Technical Report of the Japan Meteorological Agency,* No.110, 233pp.(in Japanese).

Japan Meteorological Agency, 1991. *Sankei*, 2 April. (in Japanese).

Omoto, Y. and Aono, Y., 1989. Estimation of blooming dates for *Prunus yedoensis* by means of kinematic method. *J. Agr. Meteorol.*, **45**, 25-31.

Omoto, Y. and Aono, Y., 1990. Estimation of change blooming dates of cherry flower by urban warming. *J. Agr. Meteorol.*, **46**, 123-129.

Omoto, Y. and Aono, Y., 1991. Effects of warming on changes in vegetation and phenological events. *Proc. Int. Conf. Climatic Impacts on the Environ. Soc., Jpn.* E1-6.

Oshima, H., 1987. *Kisho*, 359, 41-43. (in Japanese).

Shigehara, H., 1991. Phenological observation data in Japan to be utilized as an indicator of climatic variation. *Proc. Int. Conf. Climatic Impacts on the Environ. Soc., Jpn.*, C1-6.

Yoshino, M. and Park Ono, H.S., 1996. Variations in the plant phenology affected by global warming. (in this volume)

VARIATIONS IN THE PLANT PHENOLOGY AFFECTED BY GLOBAL WARMING

Masatoshi Yoshino[1] and Hye-Sook Park Ono[2]

ABSTRACT: In order to estimate the impact of global warming on plant phenology in Monsoon Asia, an attempt was made first to review the previous studies in Japan, Korea and China. In particular, flowering dates of cherry blossoms were analyzed in relation to air temperature in March, or as a function of latitude, longitude, coldness/warmth indices. Using the experimental formula, one can estimate the condition in the case of global warming. Secondly, it is indicated that the effects of urbanization in big cities on quickening of blooming dates are at a rate of 1-3 day par 1°C. It was shown that the mean flowering dates of cherry blossoms in Japan and Korea become 3-4 days earlier when mean air temperature of March increases 1°C. Further, in the case of early blooming the flowering dates have a more significant relation to March air temperature. In the last part of the study, differences of flowering dates in Honshu, Japan and sugar cane harvest in Minamidaito-jima, Japan, between the El Niño and La Niña events were studied. The flowering date is significantly earlier in the El Niño years. Deviations of harvested area in the El Niño years are negative for "summer planting ". The deviations are completely opposite in the La Niña years. This is caused by the dry conditions under the influence of developed zonal anticyclone in the subtropics over the North Pacific in the El Niño years, which is suggested to occur in the case of global warming.
KEY WORDS: phenology, global warming, air temperature, cherry blossom, sugar cane harvest, flowering dates, El Niño year, La Niña year, secular variation, Monsoon Asia.

1. INTRODUCTION

Plant phenology is one of the important climatological components from the viewpoint of the environment. In particular, its change in the case of global warming could provide serious impacts not only on the plant and animal life, but also on agricultural production, human health, tourism and so on. In order to consider these problems, an attempt is made first to review thoroughly our recent knowledge about the phenological events on cherry blossoms. Secondly, the changes of phenological events in East Asia due to the warming are discussed.

---

[1] Faculty of Letters, Aichi University, Toyohashi, Aichi 441, Japan.
[2] Faculty of Humanities and Social Science, Mie University, Tsu, Mie 514, Japan.

Thirdly, an example of annual change of crop cultivation in the subtropical zone is shown by the deviation of harvested area of sugar cane, as an indicator of phenological event in the lower latitudes.

2. CHERRY BLOSSOM IN JAPAN

Flowering date of cherry blossom has been studied intensively in Japan for long years (Yoshino, 1974; Kawamura, 1982), because cherry is the national flower of Japan. Monographs by Daigo (1947), Daigo and Suzuki(1947), and Nakahara (1948) were the products of the studies in Japan carried out from the agrometeorological standpoint in the first half of the 20th century. They presented maps of isolines of flowering dates in Japan and analyzed the effects of meteorological elements; mainly, monthly mean or accumulated air temperature. It was indicated that flowering dates of cherry blossom (*Prunus yedoensis*) have close relation to spring air temperature. Also, they showed experimental equations of flowering date as a linear function expressed by latitude, longitude and altitude above sea level, which has been studies as a bioclimatic law by Hopkins (1938). According to Nakahara (1948), flowering date (y) of *P. yedoensis* is expressed by

$$y = a + b(\phi - 35°) - c(\lambda - 135°) + h \tag{1}$$

where ϕ is latitude, λ is longitude and h is altitude above sea level. Since the isolines of flowering dates shift from south to north in spring, the movement looks like a frontal movement. Accordingly, the isolines are called "cherry blossom front" in Japan and they have been summarized in atlases (Daigo, 1947; Momose, 1974) and the flowering date is predicted officially by Japan Meteorological Agency in the recent years. The northward shift of isolines of mean flowering dates in Japan and Korea is shown in Fig. 1.

Because northward shifting velocity of phenological front of the early blossom is slower than the later blossom in spring (Kawamura, 1982) , distance of locations of the cherry blossom front and the ume (*plum*) flowering front, earlier occurred phenological event, is wide in Kyushu, South Japan, but is short in Hokkaido, North Japan. In other words, the difference of their flowering dates are small in Hokkaido. In some cases, they flower simultaneously.

Applying these phenomena it can be supposed that the flowering seasons of various plants, period from earlier to later flowering dates, will become longer in spring in the case of global warming.

A simplified method to estimate blooming dates of *P. yedoensis* has been proposed by using the number of days transformed to standard temperature (Aono and Omoto, 1990a, b). In their method, the starting date (D^*) was expressed as

$$D^* = 110.05 + 1.763\phi - 0.062r + 0.403\mathrm{CI} - 0.234\mathrm{WI} + 0.011\mathrm{WI}^2 \tag{2}$$

where ϕ is latitude (°N) , r is distance from coast (km), CI is coldness index (°C

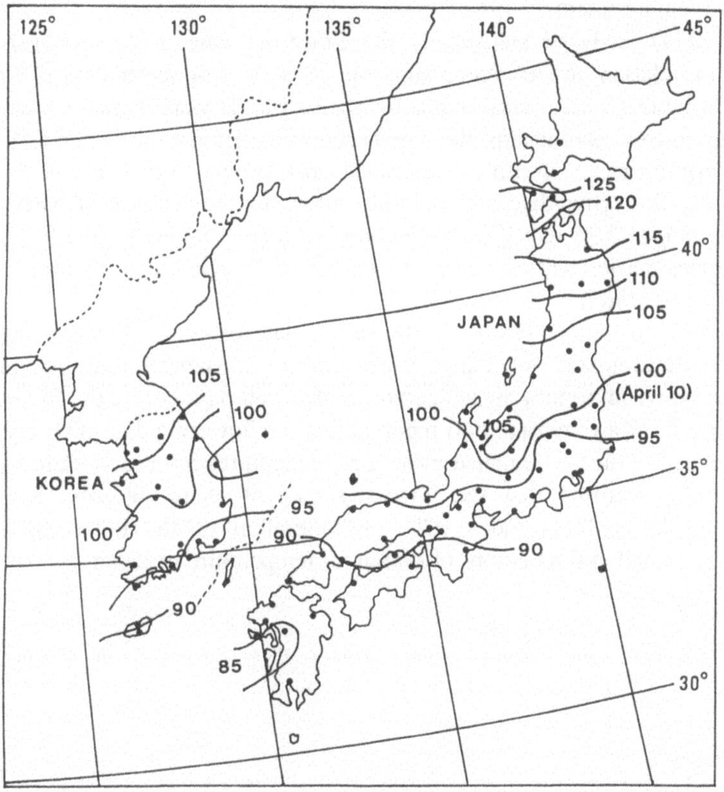

Fig. 1. Distribution of the mean flowering dates of cherry blossom (*Prunus yedoensis*) during the 37 years period, 1953-1989 (Park Ono *et al.*, 1994). Unit: Jan. 1=1, Mar.31=90 and Apr. 10=100.

month) by Kira (1945), WI is warmth index (°C month) by Kira (1945). From this experimental equation, monthly mean air temperature effects on the value of D^* in the case of warming can be obtained. According to the definition of CI/WI, the number of months with below /above 5°C of monthly mean air temperature is critical in Eq.(2). It may, therefore, result in a complicate feature from station to station, even if the warming will occur in a global scale.

Omoto and Aono (1990, 1990/91) studied the effect of urban warming on the flowering dates of *P. yedoensis*. Examples at Tokyo, Kyoto and Osaka are shown in Fig. 2. In the western part of Japan and on the pacific side of Central Japan, the dates become earlier as caused by urban warming with a rate of about 3 days/1°C. In North Japan like Aomori-City, it is the largest, but it becomes smaller in Hokkaido. The averages of earlier blooming date by urban warming at five large cities in Japan are given in Table 1. It shows that the changes of blooming date differ from city to city, but also by decade. The reason, that Nagoya shows small changes, can be attributable to the location of the observatory in Nagoya situating in a suburban environment.

Kai (1994) analyzed statistically the flowering date of *P. yedoensis* at 102 stations in relation to air temperature of 1° x 1° km mesh data. Significant relations to the CI, coldness index by Kira (1945), were found at 12 stations. According to this relationship, the date becomes earlier with a rate of 1.30 day/1°C in the warming case. Further, significant correlations were found at 61 stations between the flowering date and monthly mean air temperature of March, which imply the rate of 3.35 day/1°C in the warming case. Similarly, the full blooming date becomes earlier with a rate of 2.74 day/1°C in relation to monthly mean air temperature of March.

According to Ohno (1988), the flowering date of cherry blossom was highly related to accumulated air temperatures (mean air temperature, maximum air temperature and minimum air temperature) about 60 days before the first blooming date in Japan. Kawamura (1993) reported that the flowering date of cherry blossom was correlated to the accumulated mean air temperatures from early spring. Omasa *et al.* (1992) reported that the flowering date of cherry blossom had a good relationship to coldness index (CI) and alteration of the flowering date was intensively related to fluctuations of March air temperature in Japan.

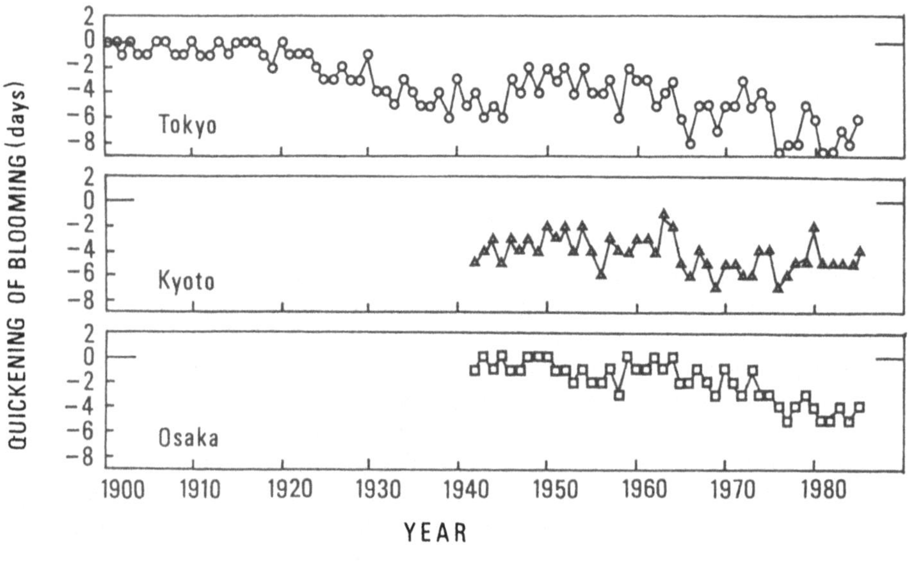

FIg. 2. Quickenings of blooming dates of *Prunus yedoensis* at three large cities (Aono, 1993).

Table 1 The centered five-year averages of change of blooming date of cherry blossoms by urban warming at selected large cities in Japan (Omoto and Aono, 1990/91). Positive means earlier.

	1965	1975	1983
Sapporo	5.2	4.2	6.2
Tokyo	5.2	6.2	7.8
Nagoya	0.2	0.2	0.8
Kyoto	3.6	5.4	4.6
Osaka	1.2	3.2	4.8

3. PHENOLOGY IN CHINA

In association with the agricultural activities in the long history of China, phenological observation and description have been made since ancient time. The earliest book of phenology as known is Xia Xiaozheng, written about 3,000 years ago. The modern phenological science has been established by Kezhen Zhu (Co-chen Chu) , 1890-1974. A review of studies on phenology in China was given by Jiang (1994) recently. According to his review, a brief description is given here in order to compare the phenological events in the respective parts of East Asia.

An earliest systematic observations in China is found in a book entitled "Diary in the years Gengzi (1180) and Xinchou (1181)" written by Lu Zhuqian, 1137-1181. This diary recorded phenological phenomena of 24 kinds of plants and animals in Jinghua District (about 29°N, 120°E, 50m a.s.l.) everyday. This is thought to be the earliest document on the phenology in the world.

Zhu and Wan (1980) summarized the phenology in China, how to observe, how to forecast and what are the relations to the date of sowing, cultivating and harvesting of some crops, based on the results of Zhu's studies for more than 50 years (Wan, 1990). In Zhu's famous paper entitled "A preliminary study on the climatic fluctuation during the last 5,000 years in China" (Zhu, 1973), it was revealed that there are good agreements between the curves reconstructed for China and δ (0^{18}) content in the ice sheet of Greenland obtained by Dansgaard *et al.* (1969), since A.D. 300. It can be concluded, therefore, that the phenological events in China will fluctuate generally in parallel with those in the Northern Hemisphere.

The other phenological studies were directed to check and to modify the Hopkin's Law (Gong *et al.*, 1983a), and to study the local climate in a small area (Zhang, 1984) and the impacts of climatic change on phenology of some woody plants in the mountainous regions.

One of the important aims of phenology from the standpoint of agricultural practice is the hybrid rice breeding. How to get good pollination of hybrid rice, namely, how to arrange the dates of sowing parent clone and of recovering clone, so that the date of blossoming of parent clone can match/meet the date of blossoming of the recovering clone. Phenological approach is necessary for further development (Ren, 1982).

Secular variation of phenological events in Beijing in spring has been reconstructed by Gong *et al.* (1983b). It was clearly shown that there were cold periods around the years 1900 and 1930, as seen for decade time scale. For the longer time scale, the period from the end of 19th century to 1950's was a cooler. It seems that the curve shows long term tendency of warming in the second half of 20th century. Although they did not discuss statistically the quickening tendency of the phenological events during the recent years, it can be roughly calculated that they are about 5 days per 100 years from 1880 to 1980, or about 4 days per 40 years from 1940 to 1980. This means warming is more striking recently.

As mentioned before, the phenological date of cherry blossom became earlier under the influence of urban warming in Japan. The rate is about 4-6 days as shown in Table 1 or about 3 days/1°C in relation to monthly mean air temperature of March in Japan. The values in Beijing in spring coincide roughly with those in Japan.

4. FLOWERING DATE OF CHERRY BLOSSOM IN EAST ASIA

In order to reexamine the shifting of flowering date of cherry blossom in East Asia, coefficients of Eq. (1) were determined by the least square method, using the data for 37 years, 1953-1989. That is,

$$y = 92.56 + 4.77\ (\phi - 35°) - 0.59\ (\lambda - 135°) + 1.28h \qquad (3)$$

The rates of difference in mean flowering dates of cherry blossom were 5 days for about 1° of latitude, 0.6 days for about 1° longitude and 1.3 days for about 100 m of altitude. In comparing with Nakahara's (1969) result, the rates of mean flowering dates of cherry blossom according to longitude and altitude were similar, but the differences according to latitude were different; i.e., his result was 6 days to 1°, but the present study's was 5 days to 1° of latitude. It can be explained that the present study covers the broader region as shown in Fig. 1 and the longer period of data.

In order to examine the relationships between the flowering date of cherry blossom and air temperature, the regression analysis was performed by using the following formula,

$$Y_j = a + bT_{ij} \qquad (4)$$

where Y_j is the flowering date of cherry blossom at j observation point (counting days calculated January 1st as one), T_{ij} is air temperature of i month at j observation point. The calculated values of a and b in Japan and Korea are shown in Table 2 respectively.

The mean flowering dates of cherry blossom have close relations to monthly mean air temperatures of winter (January) and spring temperature (March and April)

Table 2 Constant and coefficient of Eq. (4) calculated for Japan and Korea (Park Ono *et al.*, 1994)

Y_i		Japan	Korea
Y_1	a	+106.49	+98.45
	b	-3.25	-1.61
Y_2	a	+119.72	+114.44
	b	-3.57	-2.76
Y_3	a	+148.29	+163.32
	b	-4.27	-5.39

in both Japan and Korea. Especially, the mean flowering dates of cherry blossom response sensitively to mean air temperature of March. When a 1°C mean air temperature anomaly of March occurred, the flowering date of cherry blossom proceeds about 3-4 days earlier, as mentioned in previous chapter for Japan. Furthermore, in the case of early blooming, the flowering dates of cherry blossom have better relations to March mean air temperature, while in the case of later blooming, the flowering dates of cherry blossom show a better relation to mean air temperature of April.

The winter (January) and spring (March) temperatures effect on the flowering date of cherry blossom in Japan and Korea. When winter and spring temperatures were higher (HH) than mean condition (1953-1989), the flowering date of cherry blossom became earlier (a), winter and spring temperatures were lower (LL) than mean condition, the flowering date of cherry blossom were later (b), winter temperature was higher and spring temperature was lower (HL), the flowering date of cherry blossom became later (c), winter temperature was lower and spring temperature was higher (LH), the flowering date of cherry blossom became earlier (d). It can be explained that, although the flowering dates of cherry blossom were affected by not only spring air temperature but also winter air temperature, spring air temperature was more effective than winter air temperature. Especially, there is a very interest result that lower winter temperature is more effective than higher winter temperature in flowering date of cherry blossom. It might be explained that cherry blossom needs low winter temperature in a resting period before germination.

Figure 3 shows distribution of the response to the flowering date of cherry blossom against variations of March air temperature. The regionality is striking with topographical aspects in both Japan and Korea. The eastern coastal areas of the high mountain regions are larger than western part of the high mountain regions.

In Japan, when a 1°C mean air temperature anomaly occurred in March, the flowering date proceeded about 4 days earlier. However, the regional differences were obvious; e.g., Kanto and Kyusyu, through some central parts of Japan, showed a lead of above 6 days, while Tôhoku and Hokuriku showed a difference of below 4 days. In Korea, the regional differences were again obvious; e.g., Youngnam and Honam, the southern coastal parts, showed a lead of above 6 days, Hoso and

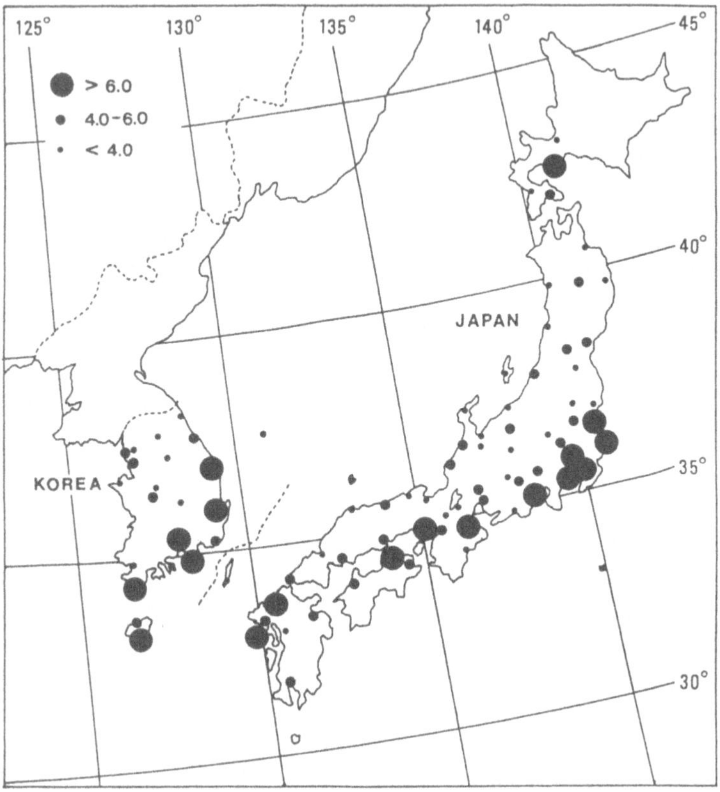

Fig. 3. Distrubution of the responses to the flowering date of cherry blossom (*Prunus yedoensis*) against variations of March air temperature (Park Ono *et al.*, 1994).

Youngso, the middle part of Korea showed about 2-3 days.

Taking into consideration on the results of similar analysis on the flowering date of cherry blossom against variations of January temperature, Japan and Korea can be divided into 3 regions based on the relationship between the flowering date and winter (January) and spring (March) air temperatures: (a) The eastern part of the high mountain areas both in Japan and Korea; e.g., Chubu, Shikoku, southeastern coastal areas of Japan, and in Yongnam and Honam, southern coastal regions of Korea, where spring temperature's effect is strong. (b) The northern areas above 36°N, where snow cover in winter is obvious, Tôhoku, the northern part, Hokuriku, the northwestern part of Japan, and in Hoso and Yongso, the middle part of Korea, where the effect of winter air temperature is strong. (c) Kanto, the middle eastern part, and Kyusyu, the southern part of Japan, and Yongdong, the middle eastern part of Korea, where the effect of air temperature is strong in spring but weak in winter.

5. SECULAR VARIATION OF FLOWERING DATE

Changes in spring air temperatures in Kyoto, Japan since the 14th century, were estimated using flowering date of cherry trees (*Prunus jamasakura*) from the old documents (Aono and Omoto, 1993, 1994). Making use of DTS method, the number of days transformed to standard temperature by Konno and Sugihara (1986), they reconstructed the variation of March air temperatures as shown in Fig. 4(b), based on the secular variation of the flowering date shown in Fig. 4(a).

If we check Fig. 4(b) in a centuries' time scale, the warm periods are from the second half of 12th to the end of 14th century (with short cold spell of mid 14th century) and from the second half of 16th to the first half of 17th century. After then, cold period prolonged from the second half of 17th to the end of 19th century. This cold period, known as "Little Ice Age", was about 2.5-3.0°C lower than present as revealed in Fig. 4(b).

Gong and Jian(1983a) reconstructed the secular variation in the periods of 1580-1710, 1711-1830 and 1831-1920. If we compare their results with the result shown in Fig. 4(b), the warmer period from the end of 16th century to the first half of 17th century and relatively warmer period among the Little Ice Age from 1740 to the first half of 19th century were found both in Japan (Kyoto) and China (Yangtze River region). Aono and Omoto (1994) reported that this period was 2°C warmer than present, estimated by the phenological data in Kyoto. Taking into consideration of these facts, it will be possible to conclude that the effects of warming tendency on phenological events in East Asia have been occurred roughly in parallel.

Okamura *et al.* (1989) reported secular variations of flowering date of cherry blossom in Japan, based on the observation at 102 stations of the Japan Meteorological Agency, from 1953 to 1988. The results were summarized for the

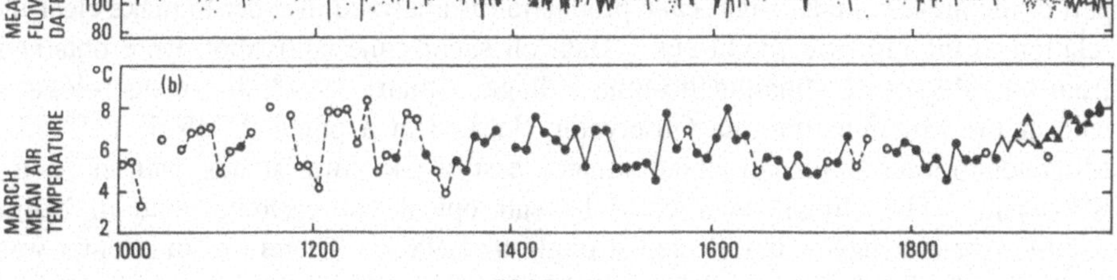

Fig. 4. (a) Secular variation of full flowering date for *P. jamasakura* at Kyoto since A.D. 1001 (dotted line shows that for *P. yedoensis*). **(b)** Variation of March means temperature since A.D. 1001, deduced from full flowering dates of cherry blossom in Kyoto city. Marks of circle and triangle show estimated values from *P. yamasakura* and *P. yedoensis* respectively. Black circle means deduced values with 5 or more years data. 10-years running means values of actual temperature since 1880's are also shown (Aono and Omoto, 1994).

13 regions, excluding 2 regions where the data were not homogeneous. Making use of their data, the flowering conditions were summarized as shown in Table 3, which presents the difference between the El Niño and La Niña events, respectively. This table shows clearly the flowering date is "early" in the El Niño event year and "late" in the La Niña event year. Using these different tendencies, it will be possible to estimate the flowering dates in the cases of global warming.

Table 3 Number of regions in Japan in terms of flowering date of cherry blossom in the case of El Niño and La Niña events

Flowering date	El Niño Events *	La Niña Events **
Very early	18	0
Early	30	5
Median (dev.-)	9	5
Median (dev.+)	5	18
Late	3	23
Very late	0	14
Total***	65	65

* El Niño event years: 1972, 1976, 1977, 1982 and 1987.
** La Niña event years: 1965, 1970, 1974, 1975 and 1988.
*** Japan is divided into 15 regions for phenological forecast at Japan Meteorol. Agency, but ,become of inhomogeneous data, 2 regions were omittted. Therefore 13 regions times 5 cases makes 65 in total.

6. SUGAR CANE HARVEST

In lower latitudes, seasonal change of phenological events become unclear. In this part of the present study, sugar cane production is analyzed in order to make clear its relation to the climatic conditions. Data on sugar cane cultivation were obtained from the "Report of Minamidaito-jima Village, January 1994", and meteorological data at the Minamidaito-jima Observatory located at a point, 25°49'N, 131°14'E. Minamdaito-jima is a small island situated east of Okinawa Islands, Nansei Shoto, SW-Japan. The climate is affected by subtropical anti-cyclonic zone in North Pacific, whose behavior is reflected strongly by ENSO. Therefore, an attempt was made to clarify deviation of harvested area of sugar cane as an indicator of phenological events in the lower latitudes, in relation to the El Niño and La Niña events.

Table 4 shows the deviation of harvested area in the years of typical El Niño and La Niña events during the last 25 years. Details for calculating deviation are given in Notes 1) - 6) of the lower part of Table 4. The results are as follows: deviation for summer planting sugar cane is negative, -15 ~ -89 ha, in the case of El Niño but positive, +11 ~ +80 ha, in the case of La Niña. Total harvested area was 1,240 ha in 1993/94. Among this total, summer planting area was 259 ha, spring planting

Table 4 Deviation of the harvested area (ha) of sugar cane in Minamidaito-jima in the cases of El Niño and La Niña events

El Niño Events				La Niña Events			
	Deviation [1] of harvested area				Deviation [1] of harvested area		
Harvest year	Summer planting[2]	Spring planting[2]	Growth from stumps	Harvest year	Summer planting[2]	Spring planting[2]	Growth from stumps
1972/73[3]	-20	+27	-33ha	1971/72	+22	-3	-31ha
1977/78	+23[a]	-6	-7	1973/74	+11	-35	+23
1982/83	-34	-38[b]	+77[c]	1976/77[4]	+33	-5	-25[d]
1987/88	-15	+58	-16	1988/89	-1[e]	-51	+53
1991/92	-89	+26	-23	1992/93[5]	-80	-47	-6
Average	-27	+26	-1	Average	+29	-28	+16
		[+25][6]	[-20][6]				

Notes: a)-e) are described in the text.
1) Deviation from 3 years running mean.
2) 15 months are needed generally from planting to harvesting.
3) For example, 1972/73 means winter of 1972/73 and spring of 1973.
4) La Niña event was 1975/76.
5) 1992/93 was not a typical La Niña event. But there was a tendency of La Niña among the long range El Niño events.
6) Average excluding values of 1982/83.

area 247 ha, and the area of growth from stumps 734 ha, respectively. So, roughly speaking, 10-15% in negative/positive deviation can be occurred in the harvested area of summer planting in the El Niño/La Niña event years. Contrastively, 10-15% of positive/negative deviation can be occurred in the harvested area of spring planting in the El Niño/La Niña event years.

There were several exceptional cases. In Table 4, notes a) - e) can be explained as follows: a) Low temperature appeared in spring 1978. So that, deviations were same to the case of La Niña. b) There were drought in 1981, and low temperature in 1982. So deviation was same to the case of La Niña. c) It is thought that this is a result of compensation of serious negative deviation of spring and summer plantings. d) Spring in 1977 was warm, although it was a La Niña year. Also, it can be explained as compensation of positive summer planting. e)There were warm and dry in 1988 and in spring 1989, although La Niña event.

Air temperature and rainfall conditions and deviation of harvested area of sugar cane in Minamidaito-jima in the case of El Niño and La Niña events are summarized in Table 5. It is of particular importance that the climatic conditions are just opposite between the El Niño event and the La Niña event. Also, deviations of harvested area are just opposite between two events. Further, it should be important that, in the case of El Niño event, higher temperature with scarce rainfall causes negative harvested area of summer planting sugar cane, but positive for spring planting. Because higher temperature and scarce rainfall are more effective on the sugar cane growth in summer, summer planting sugar cane suffers from these climate conditions. But spring planting does not suffer from them, because it has

Table 5 Air temperature and rainfall conditions and deviation of harvested area of sugar cane in Minamidaito-jima in the case of El Niño and La Niña events

Event	Average of mean ann. temperature deviation	Maximum deviation of mean ann. temperature	Annual rainfall deviation observed	Deviation of harvested area[3]		
				Summer planting	Spring planting	Growth from stumps
El Niño[1]	+0.3°C	+0.9°C	-0.56~-457mm	negative	positive	negative
La Niña[2]	-0.1°C	-0.3°C	+246~+337mm (with one excep. year)	positive	negative	positive

Notes: 1) Average of 1972, 1977, 1982,1987, 1991.
2) Average of 1971, 1973, 1976.
3) Average deviation is shown in Table 4.

already grown up stronger in summer. Growth from stumps is also negative, because of warmer and drier conditions.

In the La Niña years, the climates were cooler and wetter, which are favorable to summer planting, but not to spring planting. Growth from stumps are also favorable. From this table, we can estimate the cases in the El Niño event, which is thought to be correlated to the warming cases. Only positive deviation is expected for the spring planting sugar cane. On the other hand, negative deviation occurs in the harvested area of sugar cane summer planting and also grown from stumps, whose amount becomes to about 80% of the total cultivated area. This is caused by the higher temperature as well as severe drought conditions related to developing anticyclone zone over the subtropics of North Pacific. It may be one of the good phenological example of crop harvest, related to air temperature and rainfall change in the warming case.

7. SUMMARY AND CONCLUSION

Phenology in the case of warming in the coming century was studied first by reviewing the previous studies and secondly by the data obtained from various sources. They are summarized as follows:

(1) The flowering seasons of various plants, that is, the period from earlier to later flowering dates, will become longer in spring in the case of warming.

(2) The effect of urbanization on the flowering dates of cherry blossom becomes earlier with a rate of about 3-4 days per 1°C warming. But, it shows relatively large, local differences: large in North Japan and small in South Japan.

(3) The phenological date of cherry blossom becomes earlier under the influence of urban warming with a rate of 4-6 days per 100 years in Beijing, China, which is also found in Japanese big cities.

(4) Flowering date of cherry blossom relates closely to March air temperature, but it depends upon also winter temperature. Cold winter and warm spring are the most effective to quicken the flowering.

(5) Climatic fluctuations reconstructed by the phenological date in China coincides roughly with that in the Northern Hemisphere.
(6) In the time scale of centuries, the warm periods are from the second half of 12th to the end of 14th century and from the second half of 16th to the first half of 17th century. Little Ice Age from the second half of 17th to the end of 19th century can be seen.
(7) Such cycles were occurred in parallel with those in the regions in East Asia.
(8) Based on the facts in the El Niño event year, it can be estimated that the flowering dates in the case of warming become earlier than the present.
(9) Deviation of sugar cane harvesting area of summer planting in the El Niño event year is negative, but that of spring planting sugar cane is positive, in Minamidaito-jima, a small island in the sub-tropical zone in the North Pacific.
(10) On the contrary, it is completely opposite in the La Niña event years.
(11) Roughly speaking, 10-15% of negative/positive deviation occurs in the harvesting area of summer planting sugar cane in the El Niño/La Niña events years.
(12) Negative deviation of sugar cane harvesting area of summer planting and also grown from stumps, whose amout becomes to about 80% of total area, occurs in the case of El Niño event, associated with higher temperature and drought conditions. Similar phenomena can be estimated in the global warming.

ACKNOWLEDGMENT

Present writers should express their thanks to Prof. Jiang Ailiang, Academia Sinica, Beijing, China, who summarized the Chinese works on the theme, to Prof. Yukio Omoto, Emeritus Professor at Univ. Osaka Prefec. for presenting us their publication, and to Mrs. Keiko Kai, National Institute for Environmental Research, Tsukuba, Japan, who assisted to collect the literature. Also, our sincere thanks go to Mrs. Nobuko Toyosawa, who typed the manuscript.

REFERENCES

Aono, Y., 1993. Climatological studies on blooming of cherry tree (*Prunus yedoensis*) by mean of DTS method. *Bull. Univ. Osaka Pref., Ser. B,* Vol. **45**, 155-192.

Aono, Y. and Omoto, Y., 1990a. Estimation of blooming date for *Prunus yedoensis* using DTS combined with chill-unit accumulations. *J. Agr. Meteorol.*, **45**, 243-249. (Je)

Aono, Y. and Omoto, Y., 1990b. A simplified method for estimation of blooming date for the cherry by means of DTS. *J. Agr. Meteorol.*, **46**, 147-151. (Je)

Aono, Y. and Omoto, Y., 1993. Variation in the March temperature deduced from cherry blossom in Kyoto since the 14th century. *J. Agr. Meteorol.*, **48**, 635-638.

Aono, Y. and Omoto, Y., 1994. Estimation of temperature at Kyoto since the 11th century using flowering date of cherry trees in old documents. *J. Agr. Meteorol.*,

49, 263-272. (Je)

Daigo, Y., 1947. *Atlas of Agricultural Meteorology of Japan.* Kyoritsu Shuppan Book Co., Tokyo. 234pp.

Daigo, Y. and Suzuki , Y., 1947. *Phenology of Japan.* Hokuryukan Press, Tokyo, 187pp. (J)

Dansgaard, W. *et al.*, 1969. One thousand centuries of climate record from Camp-Century on the Greenland ice sheet. *Science*, **378**.

Gong, G. and Jian, W., 1983a. On the geographical distribution of phenodates in China. *Acta Geogr. Sinica*, **38** (1). (Ce)

Gong, G., Zhang, P. and Zhang, J., 1983b. The variations of natural phenological seasons in spring of Beijing. *Sci. Bull.*, **24**, 1517-1519. (C)

Hopkins, A.D., 1938. *Bioclimatics - A Science of Life and Climate Relations.* U. S. Dept. Agr. Miscellaneous Publ., No. 280, 187 pp.

Jiang, A., 1994. A review of studies on phenology in China. (A paper prepared to the present writer's request. Not yet published) 13 pp.

Kai, K., 1994. Kikohendo to wagakuni no seibutsu-kisetsu to no kankei (Relationship between the climatic change and phenology in Japan). A paper presented at the 20th Meeting of Japanese Study Group for WCIP and WCAP, March 1, 1994. (J)

Kai, K., Kainuma, M. and Murakoshi, N., 1995. Effects on the phenological observation of plants by global warming in Japan. (to be published in this volume)

Kawamura, T., 1982. Phenological observations and applications of the climatological analysis. *Tenki* (*Weather*) , **29**, 85-100. (J)

Kawamura, T., 1993. An attempt to reconstruct the climate in Little Ice Age using phenological data. *J. Geogr.*, **102**, 125-130. (Je)

Kira, T., 1945. Nogyochirigaku no kiso toshiteno toa no shinkikokubun (New climatic classification of East Asia as a basis of agricultural geography). Institute of Horticulture, Kyoto Univ. May Issue, 1-23. (J)

Konno, T. and Sugihara, S., 1986. Temperature index for characterizing biological activity in soil and its application to decomposition of soil organic matter. *Bull. Nat. Inst. Agroenviron. Sci.*, **1**, 51-68. (Je)

Momose, N., 1974. *Atlas of Animal and Plant Phenology of Japan.* Marunouchi Shuppan Book Co., Tokyo, 180 pp. (J)

Nakahara, M., 1948. *Phenology.* Kawadesyobo Press, Tokyo, 209 pp. (J)

Nakahara, M., 1969. Climatological study on Japanese phenology. *Reports of Department of Horticulture, Chiba University,* **17**, 89-113. (Je)

Ohno, T., 1988. A relationship between temperature and flowering date of cherry blossom - Application to reconstruction of palaeoclimate. *Undergraduate Thesis of Inst. of Geoscience*, Univ. of Tsukuba, 39pp. (J)

Okamura, T., Nakamura, N. and Shigehara, Y., 1989. Seibutsu kisetsu to kisho - Sakura no kaika ni miru - (Phenology and weather; in special reference to cherry blooming). A paper presented at the Meeting of Japanese Study Group on WCIP and WCAP, held in 1989. (J)

Omasa, K., Morita, T., Masuda, K. and Kainuma, M., 1992. Study on the effects of the plants according to the global warming. *Reports on the Global Environmental Studies* (1), Environment Agency of Japan, 307-311. (J)

Omoto, Y. and Aono, Y., 1990. Estimation of change in blooming dates of cherry flower by urban warming. *J. Agr. Meteorol.*, **46**(3), 123-129. (Je)

Omoto, Y. and Aono, Y., 1990/91. Effect of urban warming on blooming of *Prunus yedoensis*. *Energy and Buildings,* 15-16, 205-212.

Park-Ono, H.S., Kawamura, T. and Yoshino, M., 1994. Relationships between flowering date of cherry blossom (*Prunus yedoensis*) and air temperature in East Asia. *Biometeorology, pt. 2, Vol.1, Proc., 13th Intern. Congr. Biomet.*, Sept. 12-18, 1993, Calgary, Alberta, Canada, 207-220.

Ren, J., 1982. A method of calculation of the net accumulated temperatures at different leaf-ages in breeding the seeds of hybrid rice. *Agri. Meteorol.* (Beijing), **3**, 61-63. (C)

Wan, M., 1990. Dr. Zhu Kezhen's (Chu Coching's) contribution to the research works on phenology. *Acta Geogr. Sinica,* **45**, 14-21. (Ce)

Wan, M. and Liu, X., 1979. *Phenological observations of China.* Science Press, Beijing, 136pp. (C)

Yoshino, M.M., 1974. Agricultural climatology in Japan. In: *Agricultural Meteorology of Japan.* (ed. by Y. Mihara), 11-40, Univ. Tokyo Press, Tokyo.

Zhang, Fuchun, 1984. The application of phenological indicator to the climatic survey in mountainous area. *Collected Papers on the Study of Mountain Climate.* Meteorological Press, Beijing. (C)

Zhu, K., 1973. A preliminary study on the climatic fluctuations during the last 5,000 years in China. *Scientia Sinica,* **16**, 226-256.

Zhu, K. and Wan, M., 1980. *Phenology.* Science Press, Beijing, 175pp. (C)

(J: in Japanese, C: in Chinese, e: with English summary.)

USE OF A WATER TEMPERATURE-ECOLOGICAL MODEL TO SIMULATE GLOBAL WARMING EFFECTS ON A LAKE ECOSYSTEM

Masaaki Hosomi[1], Tetsu Saigusa[1], Takuya Okubo[2] and Akihiko Murakami[1]

ABSTRACT: This paper describes a newly developed combined water temperature-ecological (WT-ECO) model which is employed to simulate the effects of global warming on lake and reservoir ecosystems. The WT model includes (i) variations in the eddy diffusion coefficient based on the degree of thermal stratification and the velocity of wind, and (ii) a sub-model for simulating the freezing and thawing processes of surface water, water temperatures, and the mixing rates between two adjacent layers of water. The ECO model then uses these results to calculate the resultant effect on a lake's ecological dynamics, e.g., composition of phytoplankton species, their respective concentrations, and nutrient concentrations. When the model was benchmarked against Lake Yunoko, a dimictic lake, fairly good agreement was obtained over a 4-yr period; thereby indicating it is suitably calibrated. In addition, to assess the effects of global warming on a lake ecosystem, changes in Lake Yunoko's water temperature/quality were simulated in response to an increase in air temperature of 2 - 4 °C. Results indicate that such an increase will (i) increase thermal stratification in summer, which increases the nutrient concentrations in bottom water due to nutrient release from bottom sediment, (ii) increase the concentration of phytoplankton at the beginning of the autumn circulation period, and (iii) change the composition of phytoplankton species.
KEY WORDS: lake ecosystem, water temperature-ecological model[WT-ECO], phytoplankton species, global warming effect.

1. INTRODUCTION

It is generally known that biochemical reaction rates in freshwater ecosystems are increased by an increase in the mean air temperature and/or a decrease in the amount of annual snowfall in the watersheds, i.e., such phenomena lead to warmer water temperatures for longer times.

An increase in the water temperature of lake or reservoir systems is the most significant factor affecting their ecological balance, since the growth rate and standing crops of algae will correspondingly increase. Such conditions are

---

[1] Faculty of Engineering, Tokyo University of Agriculture and Technology, Koganei, Tokyo 184, Japan.
[2] Lake Biwa Research Institute, Otsu, Shiga 520, Japan.

undesirable because they prolong the existence of (i) dense mats of vegetation formed by the massive growth of blue-green algae such as *Microcystis* and *Anabaena* (in Japanese this phenomenon is called 'aoko'), and (ii) a musty odor produced by *Phormidium tenue* (Tsuno *et al.*, 1990). In addition, a higher water temperature expands the thermal stratification period and deepens the thermocline such that the dissolved oxygen concentration in the hypolimnion will be depleted. This in turn may possibly accelerate eutrophication as nutrients are subsequently released from bottom sediments. Hanasato *et al.* (1989) found that the higher temperatures in summer caused thermal stratification which lowered the reproduction rate of *Daphnia* by reducing its available food supply in the hypolimnion. In fact, this occurred even though a low concentration of dissolved oxygen near the bottom sediment kept predatory fish away from them.

Regarding a decrease in the amount of annual snowfall, field data from Lake Biwa, the largest lake in Japan (673.8 km^2), showed that this increases the water temperature in the hypolimnion and the standing crops of algae in the epilimnion, while decreasing the dissolved oxygen concentration in the hypolimnion (Fushimi, 1990).

Although such empirical data used with scenario analysis can provide valuable qualitative information on the effects of increased temperatures in aquatic ecosystems, this information by itself does not enable quantitative analysis of the corresponding long-term effects of global warming: an essential step in the development of models that can simulate changes in ecological responses using known increases in air temperature and rainfall/snowfall patterns.

This led to the present study which uses a newly developed ecological model of deep lakes and reservoirs in conjunction with a water temperature model in order to simulate the effects of global warming on a lake ecosystem.

2. MODEL STRUCTURE

The combined model, named the "water temperature-ecological (WT-ECO) model," is comprised of the WT model which calculates seasonal changes of water temperature versus depth in deep lakes/reservoirs, and the ECO model which calculates changes in various parameters of a lake ecosystem, e.g., composition of phytoplankton species, their respective concentrations, and nutrient concentrations. When combined, the WT model determines water temperatures and mixing rates between two adjacent layers of water, with the ECO model using these results to determine the subsequent effects on a lake's ecological dynamics.

3. WT MODEL

The WT model incorporates two main features which are described next. The first feature is that the eddy diffusion coefficient D_T is varied based on changes in (i) the degree of thermal stratification in a lake and (ii) the wind velocity over the surface of a lake. Initially, the vertical distribution of water temperature was numerically simulated using conventional models developed by Water Resources Engineers Inc. (WRE model) (1969) and Massachusetts Institute of Technology (MIT model) (Huber *et al.*, 1972). These models treated D_T as a constant. However, to suitably simulate annual changes in the vertical distribution of water temperature, D_T must be varied on a daily basis such that it reflects changing weather conditions. Therefore, by performing trial and error calculations, we found that D_T is best expressed as a function of a lake's degree of thermal stratification and the turbulence produced by the velocity of wind over its surface.

The second feature is that a sub-model is included for simulating the freezing and thawing processes of surface water. During winter in the northern parts of Japan, lakes are normally covered with a layer of ice, and consequently, each year there occurs two stratification periods (summer and winter) and two circulation periods (autumn and spring). Such a lake is called a dimictic lake. Since global warming will naturally reduce the time that a lake surface is covered with ice, which affects the water mixing conditions/temperatures, the WT model must incorporate a sub-model that can simulate the freezing and thawing processes of surface water. As the WRE and MIT models do not consider this, we developed a sub-model based on the energy balance model (Greene, 1981).

The input parameters used in the WT model are daily values of water temperature, inflow rate, and meteorological conditions, e.g., air temperature, humidity, solar radiation intensity, wind velocity, and cloud cover.

4. ECO MODEL

The employed ECO model is an improved version of our previous model (Hosomi *et al.*, 1991) as it includes a sub-model simulating nutrient exchange processes between bottom sediment and water (Hosomi and Sudo, 1992). Most conventionally employed models have treated nutrient release rates from bottom sediment using either a constant or a value determined by a forcing function (Larsen *et al.*, 1973; Imboden and Gachter, 1978; Gulliver and Stefan, 1982). These models, however, do not consider the mass balance of nutrients exchanged between bottom sediment and water, and therefore, they are inadequate for simulating the long-term effects of water temperature changes on a lake ecosystem, i.e., an increase in water temperature changes the amount of particulate organic matter that sinks and also the degradation rate of organic sediment.

Another advantageous feature of the ECO model is that it simulates changes in the composition of phytoplankton species, being a probable outcome following an increase in water temperature.

5. MODEL BENCHMARK CALCULATION

To benchmark the WT-ECO model, we used data obtained from Lake Yunoko located in Nikko National Park, Japan. This lake is a dimictic lake (mean depth =

7.4 m, surface area = 0.35 km^2) having a hydraulic retention time of about 40 d. Its inflow sources are the effluent from the Yumoto Wastewater Treatment Plant, the Odobu and Shiranesawa Rivers, and groundwater seepage, while the only outflow is the Yutaki Falls.

Lake Yunoko is known to be eutrophic due to its chlorophyll-*a* concentrations often exceeding 20 mg m^{-3} (Hosomi *et al.*, 1984, 1986). The lake's dominant species of phytoplankton are the diatoms of *Asterionella formosa*, *Synedra acus*, and *Fragilaria crotonensis* and the flagellate group of *Uroglena americana* and *Cryptomonas* sp. (Hosomi *et al.*, 1984). Based on the fact that the sinking velocity of *A. formosa* in Lake Yunoko is much higher than that of the flagellate group (Titman and Kilham, 1976; Hosomi *et al.*, 1984), and also that large diatoms are not consumed as much by *Keratella cochlearis* and *Bosmina longirostris*, which are Lake Yunoko's dominant species of zooplankton, the composition of phytoplankton species simulated by the ECO model was represented using a composition of diatoms and the flagellate group.

Lake Yunoko was selected as the benchmark for the ECO model. Input parameters were obtained using field observations and laboratory experiments on bottom sediment (Hosomi *et al.*, 1984). The Nikko Meteorological Station located near Lake Yunoko provided its daily weather conditions. Observed inflow rates/temperatures and nutrient/dissolved oxygen concentrations (Hosomi *et al.*, 1984) were extrapolated to daily-basis data. The benchmark calculation simulated the period from May 1, 1979 to December 1, 1982.

5.1 Benchmark of WT model

The simulation results by the WT model showed good correlation with observed data over the same period (Fig. 1). In particular, good agreement was obtained between the start/end times for circulation and stratification. This benchmark indicates that the WT model is suitably calibrated.

The values of parameters associated with the diffusion coefficient, being based on Lake Yunoko, were also used in the WT model when it was applied to Lake Suwa (mean depth, 4.6 m; surface area, 12.9 km^2), Lake Nojiri (mean depth, 20.8 m; surface area, 4.4 km^2), and Nagara Reservoir (mean depth, 11.9 m; surface area, 0.81 km^2). As these lakes had no lake-side meteorological station, input weather data had to be obtained from the respective nearest station. Such approximations, however, did not adversely affect the simulation because good correlation (1-2 °C) was nevertheless obtained (data not shown). It is therefore concluded that the WT model can suitably simulate vertical distributions of water temperature and associated seasonal changes in deep lakes/reservoirs having a thermal stratification period.

5.2 Benchmark of ECO model

Figures 2-4 respectively show the observed and simulated concentrations of chlorophyll-*a*, dissolved nitrogen, and dissolved oxygen in Lake Yunoko over the considered 4-yr period. Note that good correlation is obtained for the phytoplankton bloom in spring and autumn, the increase in nutrient concentrations in layers of water at the start of circulation in autumn, the depletion of nutrients in

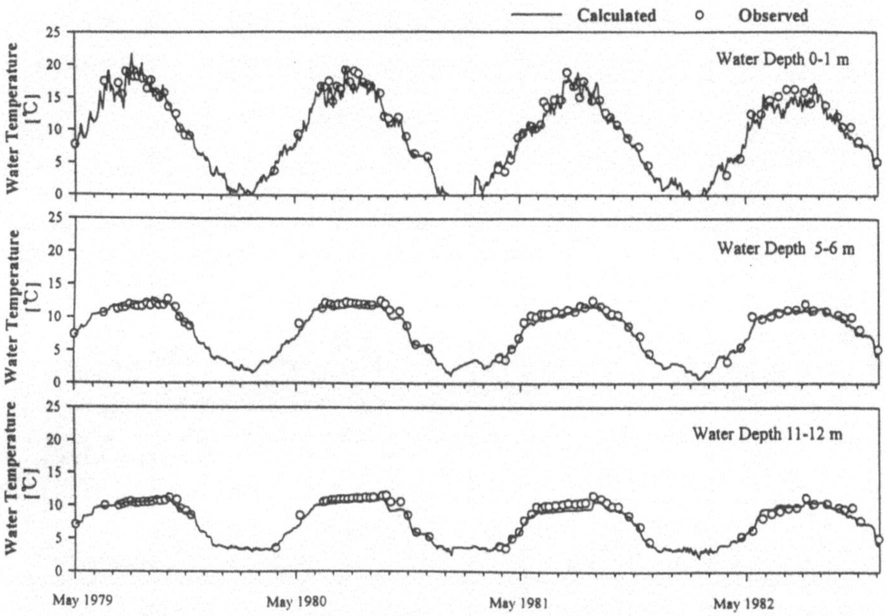

Fig. 1. Comparison of model calculations and observed data for water temperature at various water depth.

surface water and accumulation of nutrients in bottom water in summer, and the depletion of dissolved oxygen in bottom water in summer. In addition, the ECO model simulated nutrient concentrations in the bottom sediment (ca. 2.7 mg P dry sediment g^{-1}, 7.5 mg N dry sediment g^{-1}) that fell within the range of observed data. This benchmark indicates that the ECO model is suitably calibrated.

6. EFFECTS OF GLOBAL WARMING ON A LAKE ECOSYSTEM

As a benchmark calculation to assess the ability of the WT-ECO model to determine the effects of global warming on a lake ecosystem, we increased air temperature 2-4 °C higher than normal and simulated changes in Lake Yunoko's water temperatures and water quality. The temperature of inflow water was determined by the linear relationship between air temperature and water temperature. When air temperature was increased 1 °C, the inflow water from the Yumoto Wastewater Treatment Plant increased by 0.533 °C and that of the Odobu and Shiranesaw Rivers by 0.352 °C. This increase in air temperature did not increase of the temperature of groundwater seepage, which comprises more that 80% of total inflow, because no relationship is known to exist between them.

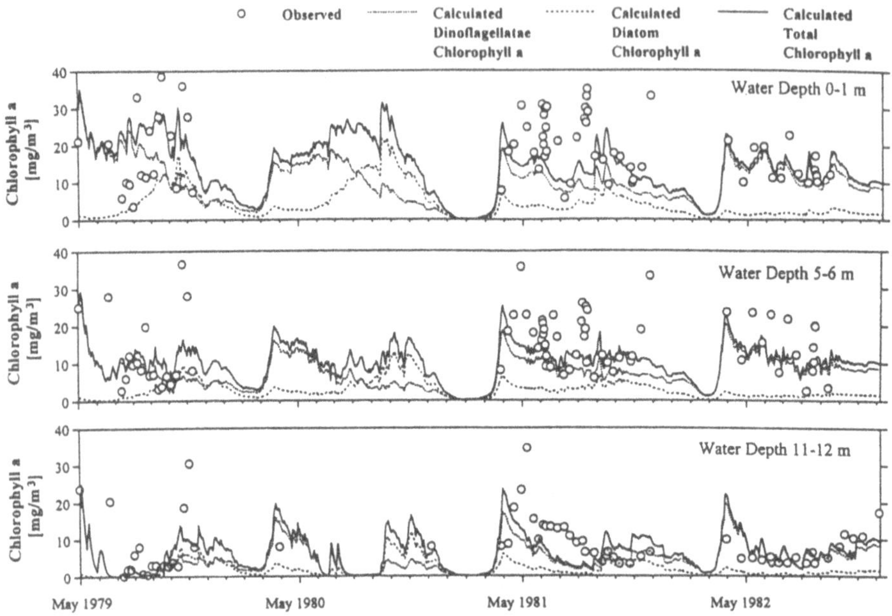

Fig. 2. Comparison of model calculations and observed data for indicated chlorophyll-*a* concentrations at various water depth.

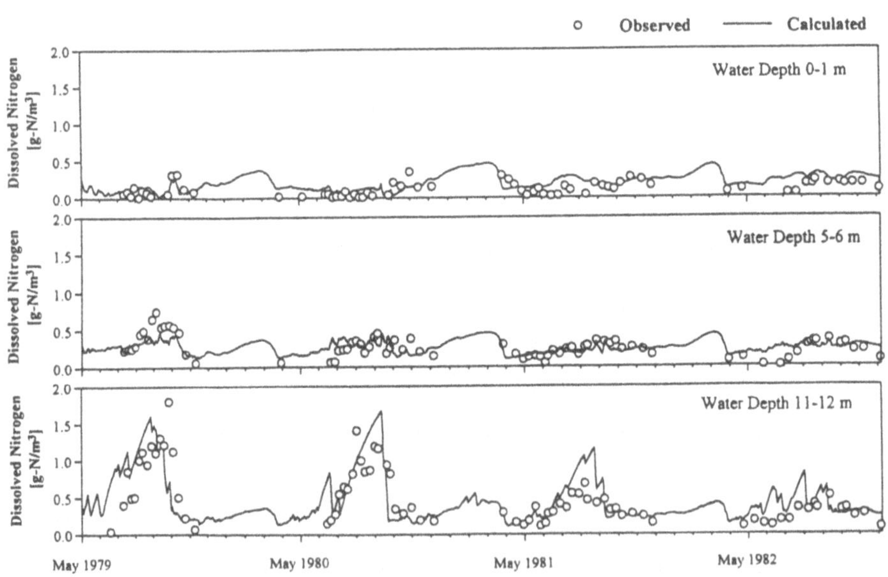

Fig. 3. Comparison of model calculations and observed data for dissolved nitrogen at various water depth.

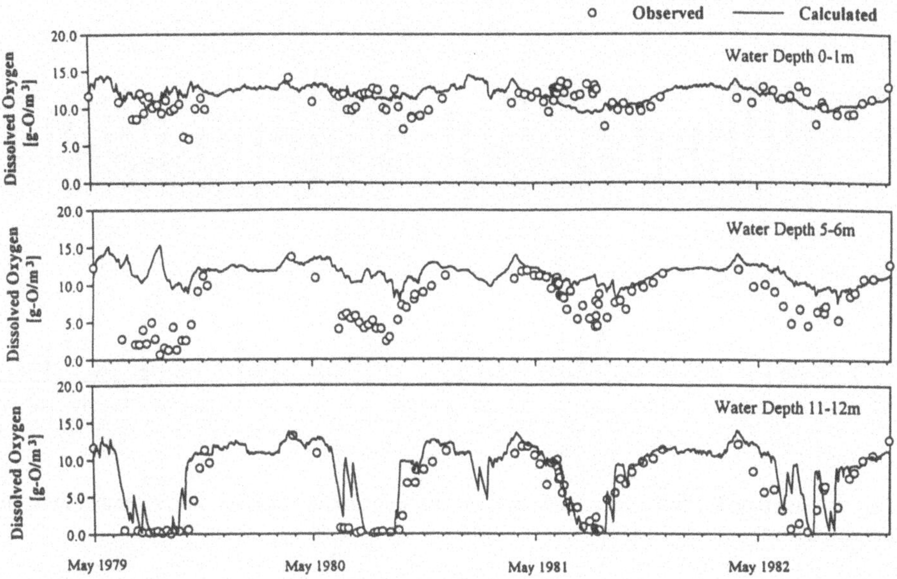

Fig. 4. Comparison of model calculations and observed data for dissolved oxygen at various water depth.

6.1 Effects on water temperature

Figure 5 shows the corresponding effect on lake water temperature. Note that the average temperature increase in surface water during the summer thermal stratification period is about 0.8 °C/°C air temperature, whereas the temperature increase of bottom water is comparatively less. During the subsequent circulation period in autumn, the increase in water temperature of all layers is about 0.5 °C/°C air temperature. These results indicate that an increase in air temperature strengthens the thermal stratification in summer due to increasing the difference between surface and bottom water temperatures.

6.2 Effects on the lake ecosystem

Figures 6 and 7 respectively show corresponding effects on chlorophyll-*a* in surface water and dissolved phosphorus in bottom water, where the strengthening of the thermal stratification in summer causes the bottom water to have an increase in the concentration of nutrients released from bottom sediment following the degradation of particulate organic matter in this sediment. In addition, when circulation of lake water begins in autumn, the accumulated nutrients in the hypolimnion are supplied to the phytoplankton in the epilimnion, which subsequently increases the concentration of phytoplankton at the start of the following circulation period (early autumn). The increase of water temperature in the surface water (0 - 1 m) also changes the composition of phytoplankton species. If the air temperature is only increased from 0 to 4 °C, the diatoms become dominant in surface water, though this

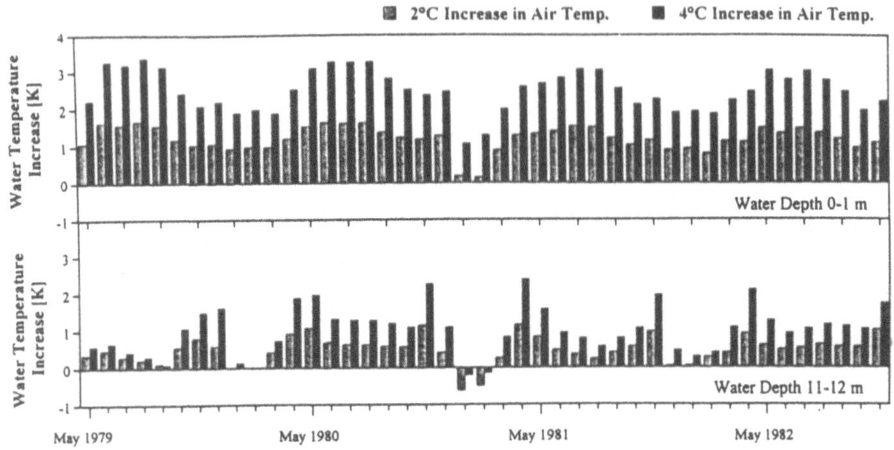

Fig. 5. Simulated effects of an increase in air temperature on the water temperature at indicated depth.

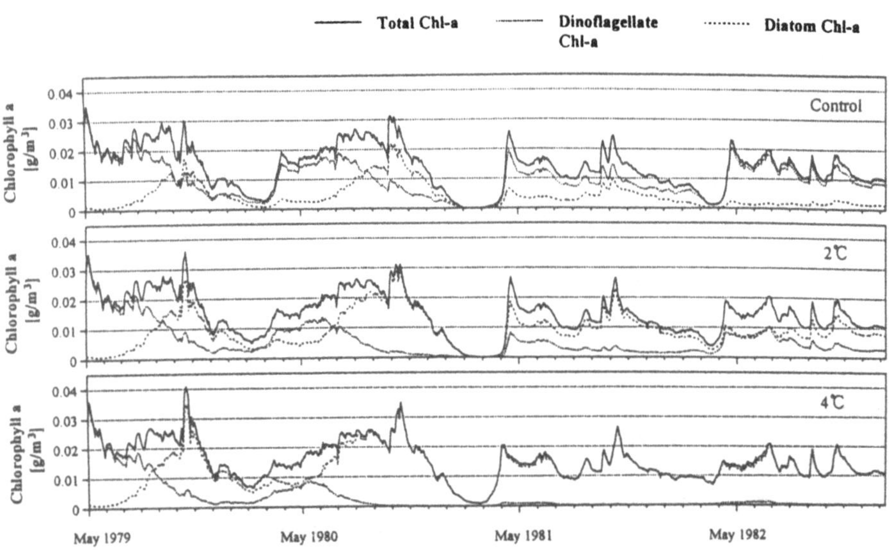

Fig. 6. Simulation of indicated chlorophyll-a concentrations for a 2 and 4 °C increase in air temperature at a water depth of 0-1 m.

affect may be lake-specific and depend on how phytoplankton growth is affected by temperature.

These simulation results indicate that an increase in air temperature will deteriorate the quality of bottom water in summer, e.g., increase the concentration of nutrients and decrease the concentration of dissolved oxygen, while also accelerating the phytoplankton bloom in autumn It should be noted that the simulation did not

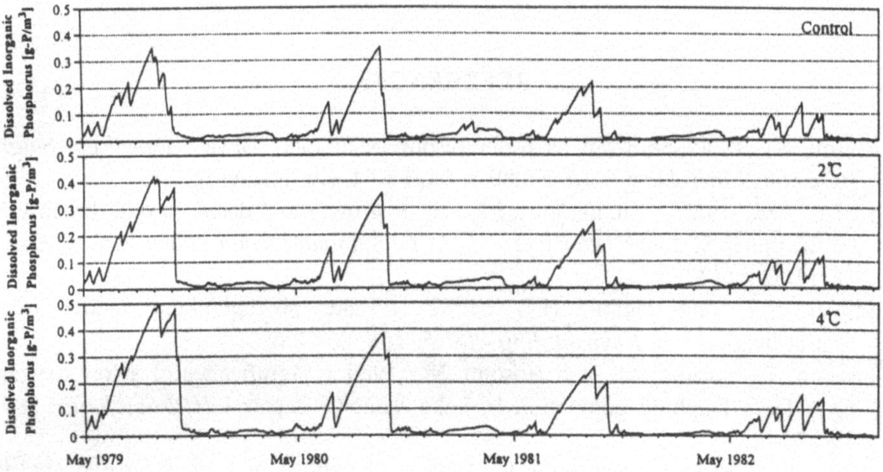

Fig. 7. Simulation of dissolved inorganic phosphorus concentration for a 2 and 4 °C increase in air temperature at a water depth of 0-1 m.

include changes in either the rainfall/snowfall patterns or the intensity of solar radiation as a result of global warming. Such changes are difficult to predict, though the former changes are important and must be considered because they affect the amount of inflow water which in turn leads to marked effects on a lake ecosystem.

7. SUMMARY AND CONCLUSION

This paper describes a newly developed combined water temperature-ecological (WT-ECO) model which is employed to simulate the effects of global warming on lake and reservoir ecosystems. The WT model simulates water temperatures and the mixing rates between two adjacent layers of water. The ECO model then uses these results to calculate the resultant effect on a lake's ecological dynamics. When the model was benchmarked against Lake Yunoko, a dimictic lake, fairly good agreement was obtained over a 4-yr period; thereby indicating it is suitably calibrated. In addition, to assess the effects of global warming on a lake ecosystem, changes in Lake Yunoko's water temperature/quality were simulated in response to an increase in air temperature of 2-4 °C. Results indicate that such an increase will (i) increase thermal stratification in summer, which increases the nutrient concentrations in bottom water due to nutrient release from bottom sediment, (ii) increase the concentration of phytoplankton at the beginning of the autumn circulation period, and (iii) change the composition of phytoplankton species. Further research will be directed at simulating rainfall/snowfall patterns as a result of global warming, and also obtaining more detailed data on the temperature-dependent characteristics of phytoplankton growth.

REFERENCES

Fushimi, K., 1990. Snowfall as water resources in Lake Biwa. *Res. Rep. Shiga Prefectural Inst. Lake Biwa,* No.89-A-01, 58-74. (in Japanese)

Greene, G.M., 1981. Simulation of ice-cover growth and decay in one dimension on the upper St. Lawrence River. *NOAA Technical Memorandum ERL CLERL-*36, PB82-114208.

Gulliver, J.S. and Stefan, H.G., 1982. Lake phytoplankton model with destratification. *J. Environ. Eng.-ASCE,* **108**, 864-882.

Hanasato, T., Yasuno, M., and Hosomi, M., 1989. Significance of a low oxygen layer for a Daphnia population in Lake Yunoko, Japan. *Hydrobiologia,* **185**, 19-27.

Hosomi, M., Okada, M. and Sudo, R., 1986. The effects of wastewater phosphorus removal on Lake Yunoko. *J. Jpn. Sewage Works Assoc.,* **23**, 52-61.(in Japanese).

Hosomi, M., Okada, M. and Sudo, R., 1991. Assessment of eutrophication control programs using an ecological model for a dimictic lake. *Water Sci. Technol.*, **24**, 339-348.

Hosomi, M., Okada, M., Yagi, O., Yamane, A. and Sudo, R., 1984. Eutrophication in Lake Yunoko. *Res. Rep. Natl. Inst. Environ. Stud., Jpn.*, No. 56, 220pp.(in Japanese)

Hosomi, M. and Sudo, R., 1992. Development of the phosphorus dynamic model in sediment-water system and assessment of eutrophication control programs. *Water Sci. Technol.*, **26**(7/8), 1981-1990.

Huber, W.C., Harleman, D.R.F. and Ryan, P.J., 1972. Temperature predication in stratified reservoirs. Proceedings of American Society of Civil Engineers. *J. Hydraulics Div.*, Paper 8839, 654-666.

Imboden, D. M. and Gachter, R., 1978. A dynamic model for trophic state prediction. *Ecol. Model.*, **4**, 77-98.

Larsen, D.P., Mercier, H.T. and Malueg, K.W., 1973. Modeling algal growth dynamics in Shagawa Lake, Minnesota, with comments concerning projected restoration of the lake. In: *Modeling the Eutrophication Process.* (eds. by E.J. Middlebrooks, D.H. Falkenborg and T.E. Maloney), Utah Water Research Laboratory, 15-32.

Titman, D. and Kilham, P., 1976. Sinking in freshwater phytoplankton; some ecological implications of cell nutrient status and physical mixing processes. *Limnol. Oceanogr.*, **21**, 409-417.

Tsuno, H., Somiya, I. and Ono, Y., 1990. Impact of global warming on urban water supply and sewerage systems. *Global Environment Research in the FY 1990.* Global Environment Department, Environment Agency of Japan, 192-194.

Water Resources Engineers, Inc., 1969. Mathematical models for perdition of thermal energy changes in impoundments. *US Environmental Protection Agency Water Pollution Control Research Series* 16130 EXT, contract 14-22422, US Government Printing Office, Washington, DC.

APPENDIX

The newly developed combined model, named the "water temperature-ecological (WT-ECO) model" is comprised of the WT model which calculates seasonal changes of water temperature versus depth in deep lakes/reservoirs, and the ECO model which calculates changes in various parameters of a lake ecosystem.

WT model

The WT model is a one-dimensional water temperature model as shown in Fig. A1. Based on water and heat energy conservation in each layer, the basic equation to define the WT model is expressed by

$$\frac{\partial T}{\partial t} = \underbrace{\frac{B}{A_i}(U_i T_{IN} - U_o T)}_{\text{inflow and outflow}} - \underbrace{\frac{1}{A}\frac{\partial}{\partial y}(V_e A T)}_{\text{vertical advection}}$$

$$+ \underbrace{\frac{1}{A}\frac{\partial}{\partial y}\left(A(\alpha_T + D_T)\frac{\partial T}{\partial y} \right)}_{\text{vertical diffusion}} - \underbrace{\frac{1}{\rho C A}\frac{\partial (A\phi)}{\partial y}}_{\text{radiation}} \qquad (1)$$

Calculation of inflow and outflow

At the inflow point inflow water is mixed and entrained with surface water. The inflow and its temperature after entrainment are expressed by

$$Q'_{IN} = (1+r)Q_{IN} \qquad \text{flow rate of inflow water} \qquad (2)$$

$$T'_{IN} = \frac{T_{IN} + T_S r}{1+r} \qquad \text{temperautre of inflow water} \qquad (3)$$

(a) Maximum horizontal velocity of inflow

$$U_{i,\max} = \frac{Q_{IN}}{\int_{y_i - d/2}^{y_i + d/2} \exp\left(-\frac{(y-y_i)^2}{2\sigma_i^2} \right) B(y)dy} \qquad (4)$$

(b) Maximum horizontal velocity of the outflow

$$U_{o,\max} = \frac{Q_{OUT}}{\int_{y_i - \delta_o/2}^{y_i + \delta_o/2} \exp\left(-\frac{(y-y_o)^2}{2\sigma_o^2}\right) B(y)dy} \tag{5}$$

(c) Standard deviation of horizontal velocity distribution of inflow

$$\sigma_i = \frac{d}{3.92} \tag{6}$$

$$d = \sqrt{\frac{Q_{IN}}{0.25B_i\sqrt{\varepsilon_i g}}} \tag{7}$$

(d) Standard deviation of horizontal velocity distribution of outflow

$$\sigma_o = \frac{\delta_o}{3.92} \tag{8}$$

$$\delta_o = \sqrt[3]{\frac{Q_{OUT}}{G\theta\sqrt{\varepsilon_o g}}} \tag{9}$$

(e) Horizontal velocity of the inflow in each layer

$$U_i(y) = U_{i,\max} \exp\left(-\frac{(y-y_i)^2}{2\sigma_i^2}\right) \tag{10}$$

(f) Horizontal velocity of the outflow in each layer

$$U_o(y) = U_{o,\max} \exp\left(-\frac{(y-y_o)^2}{2\sigma_o^2}\right) \tag{11}$$

Calculation of vertical advection

The vertical flow advected into/out of each layer can be calculated from the horizontal velocity of the inflow/outflow in each layer :

$$\frac{\partial}{\partial y}(AV_e) = B\,(U_i - U_o) \tag{12}$$

The vertical flow advected in each layer is determined by the integration of Eq.(12) from the bottom layer.

Calculation of vertical diffusion

It is essential for calculation of vertical diffusion to determine the eddy diffusion coefficient. The eddy diffusion coefficient D_T is varied based on changes in (i) the degree of thermal stratification in a lake and (ii) the wind velocity over the surface of a lake. The degree of thermal stratification is assumed to be expressed by the Richardson number Ri :

$$D_T = D_{T,S} \exp(-C_2 Ri) \tag{13}$$

$$D_{T,S} = C_1 \sqrt{C_f}\, W_{10} \tag{14}$$

$$Ri = -\frac{g\, \varepsilon\, \Delta z}{u^2} \tag{15}$$

$$C_f = (1.0 + 0.07\, W_{10}) \times 10^{-3} \tag{16}$$

$$u^2 = C_f \left(\frac{\rho_a}{\rho_S}\right) W_{10}^{\ 2} \tag{17}$$

$$\varepsilon = \frac{(\rho_S - \rho)}{\rho} \tag{18}$$

$$\rho_a = \frac{1.293}{1 + 0.00367\, T}\, \frac{H}{760} \tag{19}$$

Calculation of heat exchange

Heat exchanges through the air-water interface depend on both internal hydromechanical behavior of the water body and the physics of its interaction with the overlying air mass. The net heat energy flux across the air-water interface is expressed by the sum of the net solar radiation flux, the net atmospheric radiation flux, the water surface radiation, the evaporation heat flux, and the sensible heat flux.

(a) Net solar radiation flux

$$\phi_o = (1 - a_r)\, \phi_i \tag{20}$$

The solar radiation flux in each layer is expressed by

$$\phi = (1 - \beta)\, \phi_o e^{-\eta (y_S - y)} \tag{21}$$

(b) Evaporation heat flux and sensible heat flux

$$\phi_e + \phi_c = \frac{(3.08 + 1.85\, W_{0.15}) \times 10^{-4}}{86400}\, \rho\, (e_S - \psi\, e_a)$$

$$\times\left\{L_V + C\,T_S + \frac{269.1\,(T_S - T_a)}{e_S - \psi\, e_a}\right\} \tag{22}$$

(c) Net atmospheric radiation flux and water surface radiation flux

$$\phi_{ra} = 0.97\,k\,\left\{T_W^{\,4} - 0.937\times10^{-5}\,T_A^{\,6}\left(1+0.17\,n^2\right)\right\} \tag{23}$$

(d) Net heat energy flux across the air-water interface

$$\phi_S = \phi_o - \left(\phi_e + \phi_c + \phi_{ra}\right) \tag{24}$$

Notation

a_r : reflectivity of water surface [-]
A : area of the layer [m^2]
B : average width of the layer [m]
B_i : average width of the inflow [m]
C : specific heat of water in the layer [$kcal \cdot kg^{-1} \cdot K^{-1}$]
C_1 : constant [-]
C_2 : constant [-]
C_f : friction coefficient at the water surface [-]
D_T : eddy diffusion coefficient [$m^2 \cdot sec^{-1}$]
$D_{T,S}$: eddy diffusion coefficient at the water surface [$m^2 \cdot sec^{-1}$]
e_a : saturation vapor pressure at air temperature [mmHg]
e_S : saturation vapor pressure at the water surface temperature [mmHg]
g : gravitational acceleration [$m^2 \cdot sec^{-1}$]
G : constant : surface and bottom layer discharge = 0.324, middle discharge = 0.134
H : atmospheric pressure [mmHg]
k : Stefan-Boltzmann constant : 1.355×10^{-11} [$kcal \cdot m^{-2} \cdot sec^{-1} \cdot K^{-4}$]
L_V : latent heat of evaporation [$kcal \cdot kg^{-1}$]
n : degree of cloudiness ($0.0 < n < 1.0$) [-]
Q_{IN} : inflow rate [$m^3 \cdot sec^{-1}$]
Q'_{IN} : inflow rate after entrainment [$m^3 \cdot sec^{-1}$]
Q_{OUT} : outflow rate [$m^3 \cdot sec^{-1}$]
r : coefficient of entrainment [-]
Ri : Richardson number [-]
T : water temperature of the layer [°C]
T_a : air temperature at the water surface [°C]
T_A : air temperature at the water surface [K]
T_{IN} : water temperature of the inflow [°C]
T_S : water temperature of the surface layer [°C]
T_W : water temperature at the water surface [K]
T'_{IN} : water temperature of the inflow after entrainment [°C]
u : drift current velocity [$m \cdot sec^{-1}$]

U_i : horizontal rate of inflow [m · sec^{-1}]
$U_{i,\max}$: maximum horizontal rate of the inflow [m · sec^{-1}]
U_n : horizontal rate of outflow [m · sec^{-1}]
$U_{o,\max}$: maximum horizontal rate of the outflow [m · sec^{-1}]
V_e : vertical rate of the layer [m · sec^{-1}]
$W_{0.15}$: wind velocity at 0.15 m above the water surface [m · sec^{-1}]
W_{10} : wind velocity at 10 m above the water surface [m · sec^{-1}]
y_i : central altitude of inflow [m]
y_n : central altitude of outflow [m]
y_S : altitude of the water surface [m]
Δz : width of the layer [m]
α_T : molecular diffusion coefficient [m^2 · sec^{-1}]
β : absorption rate of the water surface [-]
ε : difference of relative density [-]
ε_i : gradient of vertical density in the inflow [kg · m^{-4}]
ε_n : gradient of vertical density in the outflow [kg · m^{-4}]
ϕ : radiation heat reaching the layer [kcal · m^{-2} · sec^{-1}]
ϕ_c : sensible heat flux [kcal · m^{-2} · sec^{-1}]
ϕ_e : evaporation heat flux [kcal · m^{-2} · sec^{-1}]
ϕ_i : solar radiation at the water surface [kcal · m^{-2} · sec^{-1}]
ϕ_n : solar radiation across the air-water interface [kcal · m^{-2} · sec^{-1}]
ϕ_{ra} : net atmospheric radiation flux and water surface radiation flux [kcal · m^{-2} · sec^{-1}]
ϕ_S : net thermal flux across the air-water interface [kcal · m^{-2} · sec^{-1}]
η : attenuation coefficient [m^{-1}]
θ : outflow angle [rad.]
ρ : density of water in the layer [kg · m^{-3}]
ρ_a : density of the air [kg · m^{-3}]
ρ_S : density of the surface layer [kg · m^{-3}]
σ_i : standard deviation in the distribution of horizontal inflow rate [m]
σ_n : standard deviation in the distribution of horizontal outflow rate [m]
ψ : relative humidity [-]

Sub-model for the WT model

The WT model incorporates a sub-model that can simulate the freezing and thawing processes of surface water. The schematic diagram of sub-model is shown in Fig. A2.

Growth of ice in surface water

It is assumed that ice with Z cm thickness is produced in surface water when surface water temperature decrease less than 0 °C. i.e.. the amount of freezing heat of ice got back to the surface water is expressed by

$$Q = 79.7\ \rho_{ICE}\ \Delta z_{ICE} \tag{25}$$

The decrease of water surface accompanied with ice production is expressed by

$$\Delta z_W = -\frac{\rho_W}{\rho_{ICE}} \Delta z_{ICE} \tag{26}$$

The change of surface water temperature due to ice production is expressed by

$$\Delta T = \frac{Q}{\rho_W \, c_W \, z_W} \tag{27}$$

Heat transfer inside ice layer

Temperature changes in ice layer depend on heat diffusion and solar radiation :

$$\frac{\partial T}{\partial t} = \frac{K}{C}\frac{\partial^2 T}{\partial z^2} + \frac{K}{C} Q_{SW} (1-a) e^{-kz} \tag{28}$$

Heat exchanges through the air-ice interface

(a) Solar radiation flux

$$Q_{nil} = Q_{rl} - \varepsilon \sigma T_s^4 \tag{29}$$

$$Q_{rl} = \left\{ \sigma T_a^4 - \left[110.6 + 5.41 \left(\sqrt{e_a} - \sqrt{e_s}\right) - 0.485E\right]\left(\frac{Q_{SW}}{Q_{SWC}}\right)^2 \right\} \times \frac{1}{86400} \tag{30}$$

(b) Sensible heat flux

$$Q_h = \rho_a C_p C_h W (T_a - T_s) \tag{31}$$

$$C_h = \frac{V_k W^*}{W \left[\ln\left(\frac{z_a}{z_0}\right) - \psi_2\right]} \tag{32}$$

$$W^* = \frac{V_k W}{\ln\left(\frac{z_a}{z_0}\right) - \psi_1} \tag{33}$$

(c) Evaporation heat flux

$$Q_{le} = \rho_a \lambda C_w W (q_a - q_s) \tag{34}$$

$$q_s = \frac{0.622\, e_s\, RH}{P - 0.378\, e_s'} \tag{35}$$

$$q_s = \frac{0.622\, e_s}{P - 0.378\, e_s} \tag{36}$$

(d) Change of ice temperature due to heat exchanges through the air-ice interface

$$\frac{\partial T}{\partial t} = \frac{Q_{ml} + Q_h + Q_{le}}{C\, \Delta z_{ICE}} \tag{37}$$

Heat transfer between ice and water
Heat transfer through the ice-water interface is expressed by

$$Q_w = h_i\,(T_w - T_m) \tag{38}$$

$$h_i = B_i\,\frac{U_w^{\;0.8}}{z_w^{\;0.2}} \tag{39}$$

The change of ice temperature due to heat transfer through the ice-water interface is expressed by

$$\frac{\partial T}{\partial t} = \frac{Q_w}{C\, \Delta z_{ICE}} \tag{40}$$

Thawing of ice layer
When ice temperature of each layer increases more than 0 °C, the ice layer is melted by heat energy supplied from the water layer just below the bottom ice layer. As reactions in the thawing processes are reverse ones in the freezing processes, the same equations in the freezing processes are used.

Notation

a : surface reflectivity [-]
B_i : empirical constant [-]
c_{ICE} : specific heat of ice [kcal · kg^{-1} · K^{-1}]
c_W : specific heat of water [kcal · kg^{-1} · K^{-1}]
C : thermal capacity : $\rho_{ICE} \times c_{ICE}$ [kcal · cm^{-3} · K^{-1}]
C_p : isobaric specific heat [kcal · kg^{-1} · K^{-1}]
e_a : saturation vapor pressure for air temperature [mmHg]
e_S : saturation vapor pressure for ice surface temperature [mmHg]
E : constant [-]
h_i : turbulent transport coefficient [kcal · m^{-2} · sec^{-1} · K^{-1}]

$$h_i = B_i \frac{U_w^{0.8}}{z_w^{0.2}}$$

k : extinction coefficient [m^{-1}]
K : thermal conductivity of ice [$kcal \cdot m^{-1} \cdot sec^{-1} \cdot K^{-1}$]
P : atmospheric pressure [mmHg]
Q : amount of freezing heat of ice [$kcal \cdot m^{-2}$]
Q_h : sensible heat flux [$kcal \cdot m^{-2} \cdot sec^{-1}$]
Q_{le} : evaporation heat flux [$kcal \cdot m^{-2} \cdot sec^{-1}$]
Q_{rl} : air radiation heat [$kcal \cdot m^{-2} \cdot sec^{-1}$]
Q_{rnl} : radiation heat [$kcal \cdot m^{-2} \cdot sec^{-1}$]
Q_{SW} : solar radiation [$kcal \cdot m^{-2} \cdot sec^{-1}$]
Q_{SWC} : maximum solar radiation [$kcal \cdot m^{-2} \cdot sec^{-1}$]
Q_w : heat transfer between ice and water [$kcal \cdot m^{-2} \cdot sec^{-1}$]
RH : relative humidity [-]
T_a : air temperature [K]
T_m : temperature at the bottom of the ice [°C]
T_S : temperature of surface ice [K]
T_w : water temperature just below the ice [°C]
ΔT : amount of surface water temperature change [K]
U_w : flow velocity of the water just below the ice [$m \cdot sec^{-1}$]
V_k : von Karman constant [-]
W : wind velocity [$m \cdot sec^{-1}$]
z_a : measurement height of wind velocity [m]
z_n : surface roughness [m]
z_w : measurement depth of flow rate [m]
z_W : thickness of surface water [m]
Δz_{ICE} : thickness of produced ice [m]
Δz_W : decrease of water surface accompanied with ice production [m]
ε : surface radiation rate [-]
λ : fusion heat of ice : 79.7 [$kcal \cdot kg^{-1}$]
ρ_a : density of air [$kg \cdot m^{-3}$]
ρ_{ICE} : density of ice at 0 °C : 0.99984×10^3 [$kg \cdot m^{-3}$]
ρ_W : density of water at 0 °C : 0.99984×10^3 [$kg \cdot m^{-3}$]
σ : Stefan-Boltzmann constant [-]

ECO model

The WT model determines water temperature, inflow and outflow rate, and mixing rate between two adjacent layers of water due to vertical advection and diffusion for each water layer. The ECO model newly developed based on the previous model can calculate changes in various parameters of a lake ecosystem using these results. The ECO model includes a sub-model simulating nutrient exchange processes between bottom sediment and water (Fig. A3).

Ecological model

A. Basic differential equations

(a) Phytoplankton

$$\frac{\partial PH_{n,i}}{\partial t} = \underset{\text{growth death}}{\left(\mu_{PH_n} - M_{PH_n}\right) PH_{n,i}} - \underset{\text{grazing}}{FIL_{PH_n} C_{g,i} PH_{n,i} ZO_i}$$

$$+ \underset{\text{sedimentation}}{SET_{PH_n} \frac{\left(PH_{n,i+1} - PH_{n,i}\right)}{H}} + \underset{\text{inflow, outflow, vertical advection and diffusion}}{F\left(PH_{n,i}\right)} \tag{41}$$

(b) Zooplankton

$$\frac{\partial ZO_i}{\partial t} = \underset{\text{grazing}}{AS_1 \left(\sum_n \left(FIL_{PH_n} PH_{n,i}\right) + \min\left(\frac{DP_i}{\sigma_{P\cdot PH}}, \frac{DN_i}{\sigma_{N\cdot PH}}\right) FIL_{DET} \right) C_{g,i}\, \sigma_{ZO\cdot PH}\, ZO_i}$$

$$\underset{\text{death}}{- M_{ZO} ZO_i} + \underset{\text{inflow, outflow, vertical advection and diffusion}}{F\left(ZO_i\right)} \tag{42}$$

(c) Dissolved inorganic phosphorus

$$\frac{\partial IP_i}{\partial t} = \underset{\text{primary production}}{-\sum_n \left(\mu_{PH_n} PH_{n,i}\right) \sigma_{P\cdot PH}} + \underset{\text{decomposition of particulate phosphorus}}{KP_i}$$

$$+ \underset{\text{excretion from grazing}}{AS_2 \left(\sum_n \left(FIL_{PH_n} PH_{n,i}\right) \sigma_{P\cdot PH} + \min\left(DP_i, DN_i \frac{\sigma_{P\cdot PH}}{\sigma_{N\cdot PH}}\right) FIL_{DET} \right) C_{g,i}\, ZO_i}$$

$$+ \underset{\text{inflow, outflow, vertical advection and diffusion}}{F\left(IP_i\right)} + \underset{\text{release from sediment}}{RELP \frac{\left(A_i - A_{i-1}\right)}{H A}} \tag{43}$$

(d) Dissolved inorganic nitrogen

$$\frac{\partial IN_i}{\partial t} = \underset{\text{primary production}}{-\sum_n \left(\mu_{PH_n} PH_{n,i}\right) \sigma_{N\cdot PH}} + \underset{\text{decomposition of particulate nitrogen}}{KN_i}$$

$$+AS_2\left(\sum_n (FIL_{PH_n}\, PH_{n,i})\, \sigma_{N\cdot PH} + \min(DP_i \frac{\sigma_{N\cdot PH}}{\sigma_{P\cdot PH}},\, DN_i)\, FIL_{DET}\right) C_{g,i}\, ZO_i$$

excretion from grazing

$$+F\,(IN_i) + RELN\,\frac{(A_i - A_{i-1})}{H\,A_i} \tag{44}$$

inflow, outflow, vertical advection and diffusion; release from sediment

(e) Particulate phosphorus

$$\frac{\partial DP_i}{\partial t} = SET_{DP}\,\frac{(DP_{n,i+1} - DP_{n,i})}{H} + \left(\sum_n (M_{PH_n}\, PH_{n,i}) + \frac{M_{ZO}\, ZO_i}{\sigma_{ZO\cdot PH}}\right) \sigma_{P\cdot PH} - KP_i$$

settlement; death of plankton; decomposition

$$+\left(1 - AS_1 - AS_2\right) \sum_n (FIL_{PH_n}\, PH_{n,i})\, C_{g,i}\, \sigma_{P\cdot PH}\, ZO_i$$

excretion from grazing

$$-\left(AS_1 + AS_2\right)\, \min(DP_i,\, DN_i \frac{\sigma_{P\cdot PH}}{\sigma_{N\cdot PH}})\, FIL_{DET}\, C_{g,i}\, ZO_i + F\,(DP_i) \tag{45}$$

grazing; inflow, outflow, vertical advection and diffusion

(f) Particulate nitrogen

$$\frac{\partial DN_i}{\partial t} = SET_{DN}\,\frac{(DN_{n,i+1} - DN_{n,i})}{H} + \left(\sum_n (M_{PH_n}\, PH_{n,i}) + \frac{M_{ZO}\, ZO_i}{\sigma_{ZO\cdot PH}}\right) \sigma_{N\cdot PH} - KN_i$$

settlement; death of plankton; decomposition

$$+\left(1 - AS_1 - AS_2\right) \sum_n (FIL_{PH_n}\, PH_{n,i})\, C_{g,i}\, \sigma_{N\cdot PH}\, ZO_i$$

excretion from grazing

$$-\left(AS_1 + AS_2\right)\, \min(DP_i \frac{\sigma_{N\cdot PH}}{\sigma_{P\cdot PH}},\, DN_i)\, FIL_{DET}\, C_{g,i}\, ZO_i + F\,(DN_i) \tag{46}$$

grazing; inflow, outflow, vertical advection and diffusion

(g) Dissolved oxygen

$$\frac{\partial DO_i}{\partial t} = \frac{(DO_S - DO_i)}{H} + \left(\sum_n (\mu_{PH_n}\, PH_{n,i}) + KR_i \sum_n PH_{n,i}\right) \sigma_{O\cdot PH} - KP_i\,\frac{\sigma_{O\cdot PH}}{\sigma_{P\cdot PH}}$$

exchange between air and water; primary production; respiration; decomposition

$$-SEDO \frac{(A_i - A_{i-1})}{H A_i} + F(DO_i) \tag{47}$$

consumption by sediment — inflow, outflow, vertical advection and diffusion

B. Adjunct equations

(a) Inflow, outflow, vertical advection and diffusion

$$F(X_i) = \frac{B_i}{A_i}\left(U_{IN,i} X_{IN,i} - U_{OUT,i} X_i\right) - \frac{1}{A_i}\frac{\partial}{\partial z}(V_e AX) - \frac{1}{A_i}\frac{\partial}{\partial z}(AE_X \frac{\partial X}{\partial z}) \tag{48}$$

inflow — outflow — vertical advection — diffusion

(b) Light intensity in water

$$I_i = I_0 \exp\left[-\int_{y_s}^{y_i}\left(a + b \sum_n PH_{n,i}\right) dh\right] \tag{49}$$

(c) Growth rate of phytoplankton

$$\mu_{PH_n} = \mu_{\max.\,PH_n}\, g(T)\, f(I)\, \frac{IP_i}{KIP_{PH_n} + IP_i}\, \frac{IN_i}{KIN_{PH_n} + IN_i} \tag{50}$$

$$g(T) = \frac{T_i}{T_{opt,\,PH_n}} \exp\left(1 - \frac{T_i}{T_{opt,\,PH_n}}\right) \tag{51}$$

$$f(I) = \frac{I_i}{IK_{PH_n}} \exp\left(1 - \frac{I_i}{IK_{PH_n}}\right) \tag{52}$$

(d) Saturation effects of zooplankton

$$C_{g,i} = \frac{k_c}{k_c + \left(\sum_n PH_{n,i} + \min(\frac{DP_i}{\sigma_{P\cdot PH}}, \frac{DN_i}{\sigma_{N\cdot PH}})\right)} \tag{53}$$

(e) Decomposition rates of particulate phosphorus and nitrogen

$$KP_i = K_P\, \theta_P^{(T_i - 20)}\, DP_i\, KDO_i \tag{54}$$

$$KN_i = K_N\, \theta_N^{(T_i - 20)}\, DN_i\, KDO_i \tag{55}$$

(f) Respiration rate of phytoplankton

$$KR_i = RES\,\theta_R^{(T_i-20)}\,KDO_i \tag{56}$$

$$KDO_i = \frac{DO_i}{KDO + DO_i} \tag{57}$$

Sediment sub-model

A. Basic differential equations

(a) Dissolved inorganic phosphorus in pore water

• Top layer of bottom sediment

$$z\,\phi_1 \frac{\partial SIP_1}{\partial t} = F_{SS}\,IPOW \frac{\phi_1}{1-\phi_1} - SIP_1\,\phi_1\,S_1 - ABS\,SIP\,z\,\phi_1 - RELP\,T$$

gain due to moving boundary; sedimentation to the 2nd layer; adsorption release

$$+D_2\,(\phi_1+\phi_2)\,\frac{(SIP_2 - SIP_1)}{2\,z} + \rho\,(1-\phi_1)\,z\,GP_1 \tag{58}$$

diffusion; decomposition of particulate phosphorus

• Sediment layers other than top layer of bottom sediment

$$z\,\phi_i \frac{\partial SIP_i}{\partial t} = SIP_{i-1}\,\phi_{i-1}\,S_{i-1} - SIP_i\,\phi_i\,S_i - D_i\,(\phi_{i-1}+\phi_i)\,\frac{(SIP_i - SIP_{i-1})}{2\,z}$$

sedimentation; diffusion across the upper interface

$$+\,D_{i+1}\,(\phi_i+\phi_{i+1})\,\frac{(SIP_i - SIP_{i+1})}{2\,z} + \rho\,(1-\phi_i)\,z\,GP_i \tag{59}$$

diffusion across the lower interface; decomposition of particulate phosphorus

(b) Dissolved inorganic nitrogen in pore water

• Top layer of bottom sediment

$$z\,\phi_1 \frac{\partial SIN_1}{\partial t} = F_{SS}\,INOW \frac{\phi_1}{1-\phi_1} - SIN_1\,\phi_1\,S_1 - RELNT$$

gain due to moving boundary; sedimentation to the 2nd layer; release

$$+\,D_2\,(\phi_1+\phi_2)\,\frac{(SIN_2 - SIN_1)}{2\,z} + \rho\,(1-\phi_1)\,z\,GN_1 \tag{60}$$

diffusion; decomposition of particulate nitrogen

• Sediment layers other than top layer of bottom sediment

$$z\,\phi_i\,\frac{\partial SIN_i}{\partial t} = \underbrace{SIN_{i-1}\,\phi_{i-1}\,S_{i-1} - SIN_i\,\phi_i\,S_i}_{\text{sedimentation}} - \underbrace{D_i\,(\phi_{i-1}+\phi_i)\,\frac{(SIN_i - SIN_{i-1})}{2\,z}}_{\text{diffusion across the upper interface}}$$

$$+ \underbrace{D_{i+1}\,(\phi_i+\phi_{i+1})\,\frac{(SIN_i - SIN_{i+1})}{2\,z}}_{\text{diffusion across the lower interface}} + \underbrace{\rho\,(1-\phi_i)\,z\,GN_i}_{\text{decomposition of particulate phosphorus}} \tag{61}$$

(c) Particulate phosphorus in bottom sediment

• Top layer of bottom sediment

$$z\,\rho\,(1-\phi_1)\,\frac{\partial SDP_1}{\partial t} = \underbrace{F_{SS}\,DPOW\,\frac{\phi_1}{1-\phi_1}}_{\text{gain due to moving boundary}} - \underbrace{SETL_{DP}}_{\text{particulate phosphorus deposition from lake water}} - \underbrace{SDP_1\,\rho\,(1-\phi_1)\,S_1}_{\text{sedimentation to the 2nd layer}}$$

$$+ \underbrace{ABS\,SIP_1\,z\,\phi_1}_{\text{adsorption}} - \underbrace{GP_1\,z\,\rho\,(1-\phi_1)}_{\text{decomposition}} \tag{62}$$

• Sediment layers other than top layer of bottom sediment

$$z\,\rho\,(1-\phi_i)\,\frac{\partial SDP_i}{\partial t} = \underbrace{SDP_{i-1}\,\rho\,(1-\phi_{i-1})\,S_{i-1}}_{\text{sedimentation from the upper layer}} - \underbrace{SDP_i\,\rho\,(1-\phi_i)\,S_i}_{\text{sedimentation to the lower layer}}$$

$$\underbrace{-GP_i z\rho(1-\phi_i)}_{\text{decomposition}} \tag{63}$$

(d) Particulate nitrogen in bottom sediment

• Top layer of bottom sediment

$$z\,\rho\,(1-\phi_1)\,\frac{\partial SDN_1}{\partial t} = \underbrace{F_{SS}\,DNOW\,\frac{\phi_1}{1-\phi_1}}_{\text{gain due to moving boundary}} - \underbrace{SETL_{DN}}_{\text{particulate phosphorus deposition from lake water}}$$

$$\underbrace{-SDN_1\,\rho\,(1-\phi_1)\,S_1}_{\text{sedimentation to the 2nd layer}} - \underbrace{GN_1\,z\,\rho\,(1-\phi_1)}_{\text{decomposition}} \tag{64}$$

• Sediment layers other than top layer of bottom sediment

$$z\,\rho\,(1-\phi_i)\frac{\partial\, SDN_i}{\partial\, t} = \underbrace{SDN_{i-1}\,\rho\,(1-\phi_{i-1})\,S_{i-1}}_{\text{sedimentation from the upper layer}} - \underbrace{SDN_i\,\rho\,(1-\phi_i)\,S_i}_{\text{sedimentation to the lower layer}}$$

$$-\underbrace{GN_i\,z\,\rho\,(1-\phi_i)}_{\text{decomposition}} \tag{65}$$

B. Adjunct equations

(a) Diffusion coefficient in the *i*th sediment layer

$$D_i = \phi_i^2\,DIF \tag{66}$$

(b) Adsorption rate

$$ABS = ABS_0 \sum_i \frac{DO_i}{KDO + DO_i} \tag{67}$$

(c) Sedimentation rate in the *i*th sediment layer

$$S_i = \frac{F_{SS}}{1-\phi_i} \tag{68}$$

(d) Particulate phosphorus deposition from lake water to top layer of sediment

$$SETL_{DP} = \sum_i \left\{ \left(\sum_n (SET_{PH_n}\,PH_{n,i})\,\sigma_{P\cdot PH} + SET_{DP}\,DP_i \right) \frac{A_i - A_{i-1}}{A_s} \right\} \tag{69}$$

(e) Particulate nitrogen deposition from lake water to top layer of sediment

$$SETL_{DN} = \sum_i \left\{ \left(\sum_n (SET_{PH_n}\,PH_{n,i})\,\sigma_{N\cdot PH} + SET_{DN}\,DN_i \right) \frac{A_i - A_{i-1}}{A_s} \right\} \tag{70}$$

(f) Decomposition rate of particulate phosphorus in the *i*th sediment layer

$$GP_i = (SDP_i - SEDP_0)\,KSED\,\exp(-\alpha_i)\,\theta_{SED}^{\;T-20} \tag{71}$$

(g) Decomposition rate of particulate nitrogen in the *i*th sediment layer

$$GN_i = (SDN_i - SEDN_0)\,KSED\,\exp(-\alpha_i)\,\theta_{SED}^{\;T-20} \tag{72}$$

(h) Release rate of phosphorus from bottom sediment

$$RELP_i = D_1\ (1+\phi_1)\ \frac{SIP_1 - IP_i}{z} \tag{73}$$

$$RELPT = \sum_i RELP_i \tag{74}$$

(i) Release rate of nitrogen from bottom sediment

$$RELN_i = D_1\ (1+\phi_1)\ \frac{SIN_1 - IN_i}{z} \tag{75}$$

$$RELNT = \sum_i RELN_i \tag{76}$$

Notation

a : extinction coefficient by lake water [m^{-1}]
A_i : area in the ith layer [m^2]
ABS_0 : adsorption rate constant in sediment [day^{-1}]
AS_1 : assimilation rate of zooplankton [-]
AS_2 : excretion rate of dissolved inorganic nutrient with grazing [-]
b : specific extinction coefficient of phytoplankton [$m^2 \cdot g\ Chl.a^{-1}$]
B_i : width in the ith layer [m]
DIF : diffusion coefficient in sediment [$m^{-2} \cdot day^{-1}$]
DN_i : concentration of particulate nitrogen in the ith layer [$g\ N \cdot m^{-3}$]
DO_i : concentration of dissolved oxygen in the ith layer [$g\ O \cdot m^{-3}$]
DO_S : saturated concentration of dissolved oxygen [$g\ O \cdot m^{-3}$]
DP_i : concentration of particulate phosphorus in the ith layer [$g\ P \cdot m^{-3}$]
E_X : diffusion coefficient [$m^2 \cdot day^{-1}$]
FIL_{PHn} : maximum filtration rate of zooplankton for phytoplankton species "n" [$m^3 \cdot g\ C^{-1} \cdot day^{-1}$]
FIL_{DET} : maximum filtration rate of zooplankton for suspended detritus [$m^3 \cdot g\ C^{-1} \cdot day^{-1}$]
F_{SS} : sedimentation flux of particulate matter [$g\ dry \cdot m^{-2} \cdot day^{-1}$]
H : thickness of each layer [m]
I_o : solar radiation under the water surface [$kcal \cdot m^{-2} \cdot day^{-1}$]
IK_{PHn} : optimum light intensity for phytoplankton species "n" [$kcal \cdot m^{-2} \cdot day^{-1}$]
IN_i : concentration of dissolved inorganic nitrogen in the ith layer [$g\ N \cdot m^{-3}$]
$INOW$: average concentration of dissolved inorganic nitrogen in the overlying lake water of sediment surface [$g\ N \cdot m^{-3}$]
IP_i : concentration of dissolved inorganic phosphorus in the ith layer [$g\ P \cdot m^{-3}$]
$IPOW$: average concentration of dissolved inorganic phosphorus in the overlying lake water of sediment surface [$g\ P \cdot m^{-3}$]

k_c : half-saturation coefficient for saturating phenomenon [g Chl.a · m^{-3}]
K_{exc} : exchange rate between air and water [m · day^{-1}]
K_N : decomposition rate of particulate nitrogen at 20 °C [day^{-1}]
K_P : decomposition rate of particulate phosphorus at 20 °C [day^{-1}]
KDO : half saturation constant for consumption of dissolved oxygen [g O · m^{-3}]
KIN_{PHn} : half-saturation constant for dissolved inorganic nitrogen uptake of phytoplankton species "n" [g P · m^{-3}]
KIP_{PHn} : half-saturation constant for dissolved inorganic phosphorus uptake of phytoplankton species "n" [g P · m^{-3}]
$KSED$: decomposition rate of detritus in sediment at 20 °C [day^{-1}]
M_{PHn} : mortality of phytoplankton species "n" [day^{-1}]
M_{ZO} : mortality of zooplankton [day^{-1}]
$PH_{n,i}$: concentration of phytoplankton species "n" in the ith layer [g Chl.a · m^{-3}]
RES : respiration rate of phytoplankton at 20 °C [day^{-1}]
SDN_i : particulate nitrogen concentration in the ith sediment layer [g N · g dry^{-1}]
SDP_i : particulate phosphorus concentration in the ith sediment layer [g P · g dry^{-1}]
$SEDN_0$: refractory nitrogen content in sediment [g N · g dry^{-1}]
$SEDO$: oxygen consumption rate of sediment [g O · m^{-2} · day^{-1}]
$SEDP_0$: refractory phosphorus content in sediment [g P · g dry^{-1}]
SET_{DN} : settling rate of particulate nitrogen [m · day^{-1}]
SET_{DP} : settling rate of particulate phosphorus [m · day^{-1}]
SET_{PHn} : settling rate of phytoplankton species "n" [m · day^{-1}]
SIP_i : concentration of dissolved inorganic phosphorus in pore water in the ith sediment layer [g P · m^{-3}]
SIN_i : concentration of dissolved inorganic nitrogen in pore water in the ith sediment layer [g N · m^{-3}]
T_i : water temperature in the ith layer [°C]
$T_{opt,PHn}$: optimum temperature for phytoplankton species "n" [°C]
$U_{IN,i}$: inflow rate in the ith layer [m · day^{-1}]
$U_{OUT,i}$: outflow rate in the ith layer [m · day^{-1}]
V_e : vertical advected flow rate in the ith layer [m · day^{-1}]
$X_{IN,i}$: inflow concentration of item "X" [g · m^{-3}]
Z : thickness of each sediment layer [m]
ZO_i : concentration of zooplankton in the ith layer [g C · m^{-3}]
α : attenuation coefficient of decomposition rate along depth of sediment [m^{-1}]
ϕ_i : porosity in the ith sediment layer [-]
$\mu_{max \cdot PHn}$: maximum specific growth rate of phytoplankton species "n" [day^{-1}]
θ_N : temperature coefficient for decomposition of particulate nitrogen [-]
θ_P : temperature coefficient for decomposition of particulate phosphorus[-]
θ_R : temperature coefficient for respiration of phytoplankton [-]
θ_{SED} : temperature coefficient for decomposition rate in sediment [-]

ρ : density of sediment particle [g dry · m^{-3}]
$\sigma_{N\cdot PH}$: nitrogen to chlorophyll-a ratio [g N · g Chl.a^{-1}]
$\sigma_{O\cdot PH}$: oxygen to chlorophyll-a ratio [g O · g Chl.a^{-1}]
$\sigma_{P\cdot PH}$: phosphorus to chlorophyll-a ratio [g P · g Chl.a^{-1}]
$\sigma_{ZO\cdot PH}$: carbon to chlorophyll-a ratio [g C · g Chl.a^{-1}]

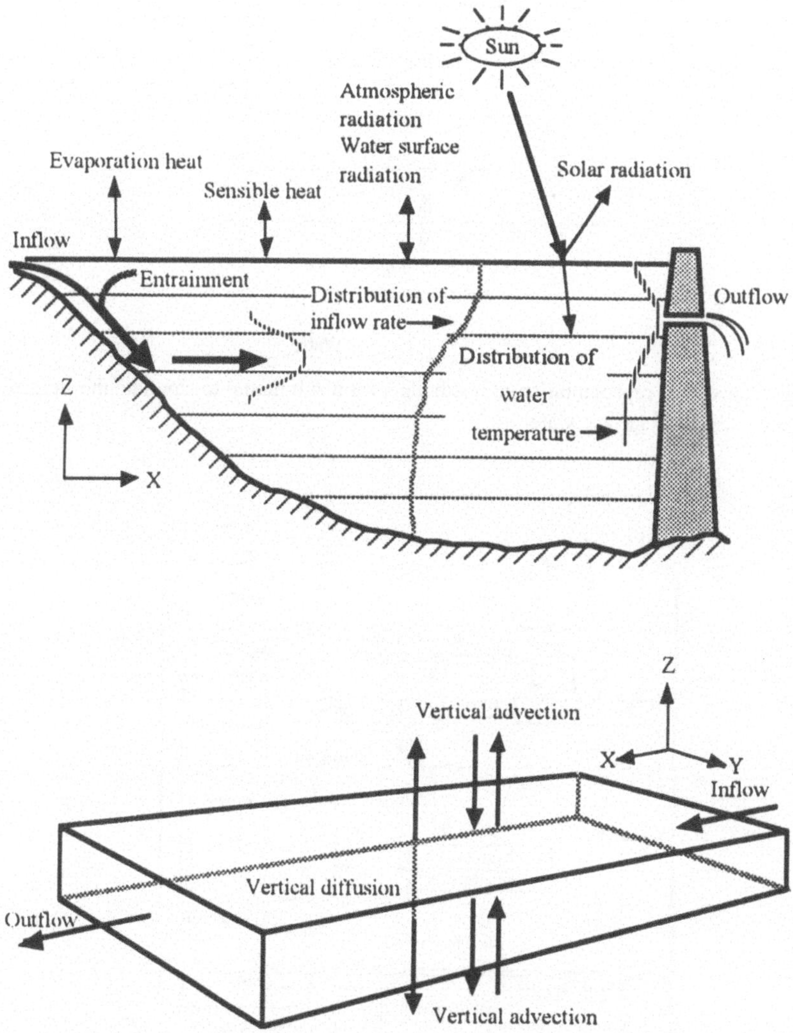

Fig. A1. Schematic diagram of the WT model.

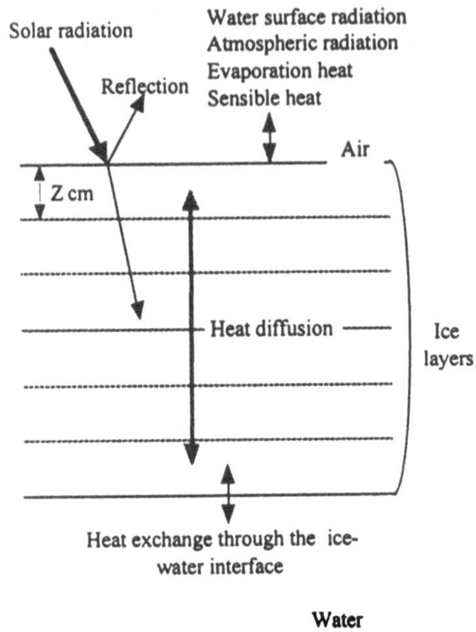

Fig. A2. Conceptual representation of heatbudget for a sub-model to simulate the freezing and thawing processes of surface water.

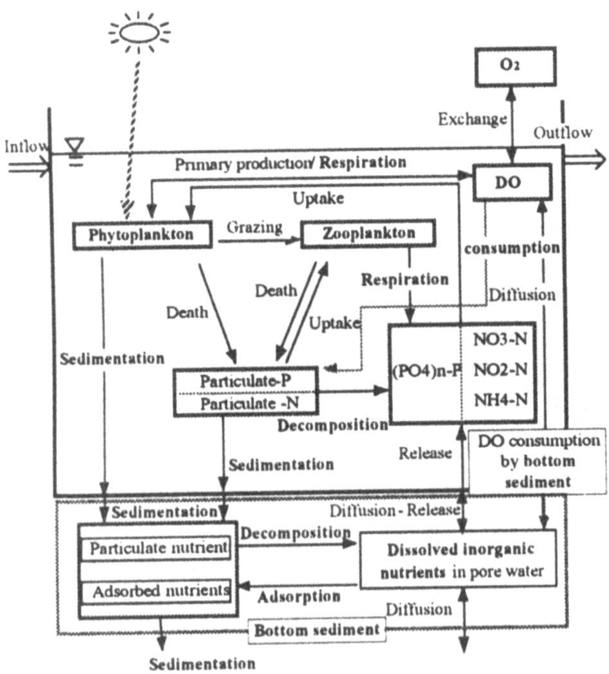

Fig. A3. Schematic diagram of the ECO model.

Ⅱ. Mountain Vegetation

CLIMATIC CHANGE AND SNOWPATCHES, REVEALED BY SOIL STRATIGRAPHY AROUND THE NIVATION HOLLOWS

Hiromu Daimaru[1] and Shigeto Ikeda[1]

ABSTRACT: This study tries to reconstruct past climatic conditions by means of alpine-subalpine meadow soil stratigraphy, which records past snow disappearance time in nivation hollows. Buried peat found on Mt. Zarumori indicates that the date of snow cover disappearance in approximately 1000 y.B.P., was earlier than in recent times. Calculations of the past snowmelt rate by the degree-day method, indicate that warming in summer and lower winter snowfall accumulation, hastened snowmelt and enabled the peat to deposit. This warmer climatic condition in north Japan corresponds to the so-called "Medieval Warm Period" of Europe and North America.
KEY WORDS: climatic change, snow patches, nivation hollows, north Japan.

1. INTRODUCTION

There are no active glaciers in Japan, so it has been very difficult to discuss Holocene climatic change by glaciological evidence. Therefore, we tried to use ephemeral snow patches as a climatic indicator in order to reconstruct the past climate. The northeastern part of Japan has one of the heaviest annual snowfall accumulations in the world. In the subalpine zone, much snow deposition (often exceeding 10 m) occurs on the southeastern slopes of mountain ridges in winter. In summer, this forms isolated snow patches, which often fill active or relict nivation hollows (Koizumi, 1982).

Many ecologists have recognized that snowmelt gradient makes a clear zonation of vegetation and soil in alpine meadows (Billings and Bliss, 1959; Stanton *et al.*, 1994). Although the relationship of summer snow patch size to climate changes for several years has been discussed (Higuchi *et al.*, 1979; Watson *et al.*, 1994), the long-term change of snow patch size since Holocene is not mentioned. In Japan, some geographers and pedologists have reported the historical change of surface soils in alpine-subalpine meadows or nivation hollows, using ^{14}C dating and tephrochronology of buried peaty soils (Osumi and Kumada, 1971; Koizumi, 1982; Yamanaka, 1983; Kariya, 1994).

In the subalpine zone of Japan, meadows with peaty soils are found where snow disappears by mid July, and sparse grasslands or moss fields with mineral-rich soil

[1] Tohoku Research Center, Forestry and Forest Products Research Institute, 72 Nabeyashiki Shimokuriyagawa, Morioka, Iwate 020-01, Japan.

will be found where snow disappears after late July. This is because the plant growth period is too short to product biomass, which is neccessary in order to form peat in such lately, snowreleased places (Yamanaka, 1983). Due to much eolian dust fall (Mizota and Matsuhisa, 1985) or volcanic ash fall, alpine-subalpine meadows in Oou Mountains usually have thick surface soils which contains good staratigraphic records, often with key-tephras.

If climatic conditions change, then the snowmelt rate will change, as will the vegetational and soil zones in the meadows. If such changes are recorded by making soil profiles in aggradational parts of the meadow, the stratigraphic sequence of the surface soil will show the historical change of snow cover disappearance time or the snow patch's size in summer. Most of datings of buried peat collected from many Japanese alpine-subalpine meadows or nivation hollows are thought to be formed in the Hypsithermal warming period (Osumi and Kumada, 1971; Yamanaka, 1983; Kariya, 1994), the warmest epoch during the Holocene. Such buried peat layers indicate that soil zonation has shifted under past, warmer climatic conditions. We, however, have no quantitative studies about climatic change and snowmelt rate for buried peat deposition.

We found a series of fossil peaty soils dated ca. 1000 y.B.P. in a subalpine meadow of northeastern Japan. The past climatic conditions which enabled the buried peat to deposit, was estimated by the degree-day method, the simplest way to calculate the snowmelt rate under changing temperatures.

2. STUDY AREAS

The studied meadow is on the southeast slope of Mt. Zarumori (1,541 m) in the Oou Mountain Range, which runs through northern Honshu, Japan (Fig. 1, Color Plates 40 and 41). Mt. Zarumori is an old volcano and consists of basaltic andesite lava, and has a steep slope on its northwest side, and a gentle slope with shallow depressions, (probably Pleistocene fossil nivation hollows) on its southeast side.

Based on 17 years of records (1936 to 1952) from Morioka (155 m a.s.l.) and Mt. Iwate (2,000 m a.s.l) (Morioka Local Meteorological Observatory, 1966.), the mean annual temperature of Mt. Zarumori is estimated as being 0.9° C with a mean of 14.5° C for the warmest month of July, and a mean of -12.6° C for the coldest month of January. The summit of Mt. Zarumori is not so high, located in the subalpine temperature zone. However, it has severe snowstorms in winter, so the mountaintop lacks trees. Furthermore, a snowbank forms each winter on the meadow, which has a gradient of approximately 10-20°, with a conspicuous break line on its leeside end (Color Plate 40).

In northeastern Japan, a white pumiceous volcanic ash called To-a (Towada-a), fell in 915 A.D. (Fig. 1, Color Plate 42; Japan Association for Quaternary Research, 1987). This key ash bed is very important for studying the paleoenvironment of this region.

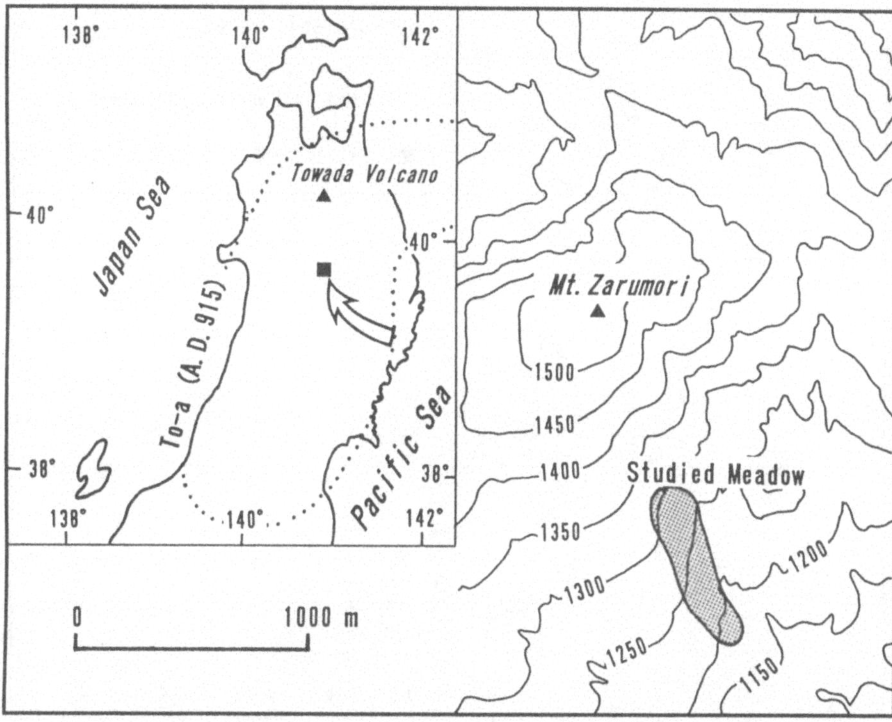

Fig. 1. Study Area. The dotted line shows the distribution of To-a tephra (after Japan Association for Quaternary Research, 1987).

3. RESULTS

At Mt. Zarumori, buried peat was found under the mineral-rich surface soil (Fig. 2, Color Plate 42). To-a (erupted in 915 A.D.) was found in the lower part of the peat layer.

This indicates that the buried peat was deposited before 915 A.D. and buried several hundreds years later. Moreover, the many soil profiles taken indicate that the surface peat layer began to deposit about 1000 y.B.P. (Fig. 2).

Many climatologists have suggested that the climate from the ninth to the fourteenth centuries, was warmer than that of the past three hundred years. This epoch is called the "Medieval Warm Period" in Europe (Hughes, 1994; Lamb, 1977; Yoshino, 1978). The colder climate of the seventeenth to the nineteenth centuries, is known as the "Little Ice Age".

The buried peat at Mt. Zarumori is thought to have formed during the Medieval Warm Period, when the snow melted earlier than today, and the snow free period for plant growth was long enough for peat deposition.

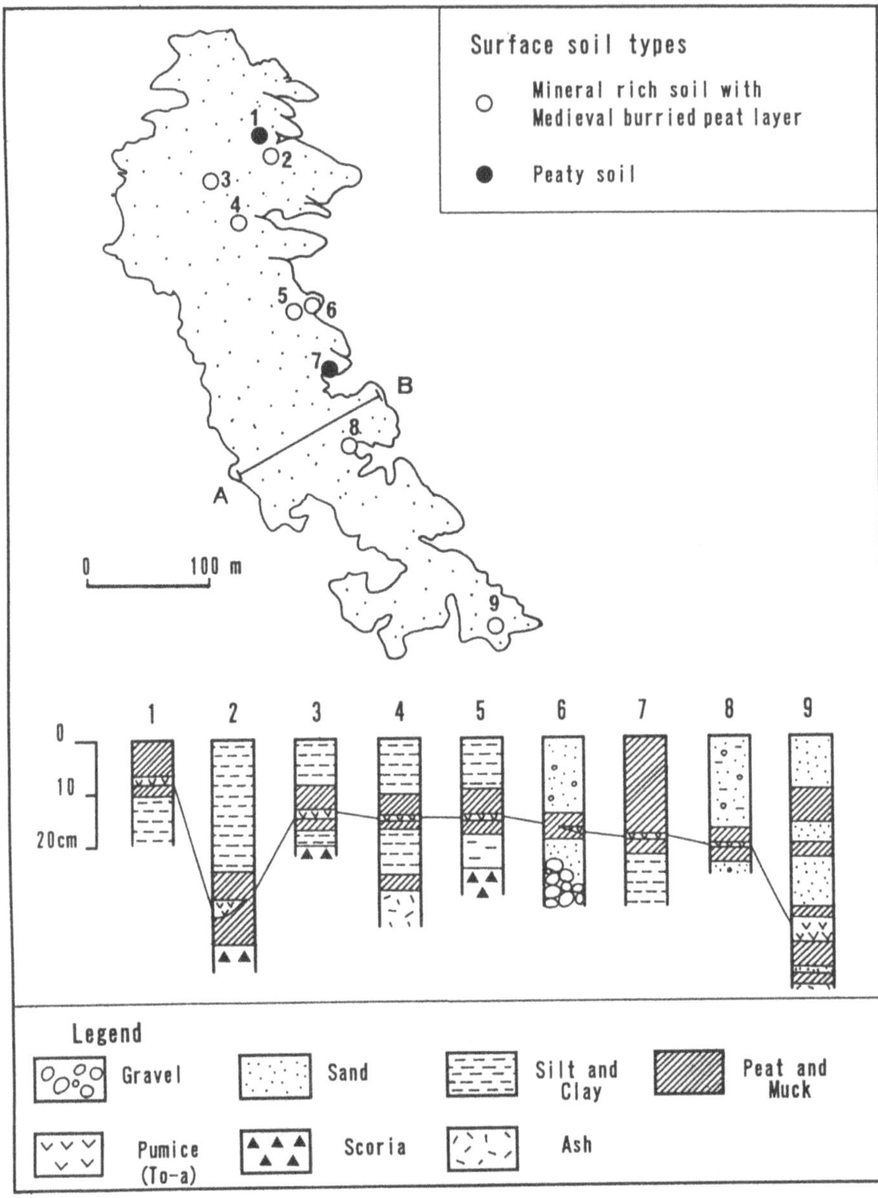

Fig. 2. Distribution of buried peaty soil in the studied meadow. The shape of the meadow was copied from an aerial photograph.

4. DISCUSSION

How much must the summer temperature and snowmelt rate change, in order to affect the zoning of soil types in nivation hollows? The degree-day method provides the best and simplest way to evaluate the snowmelt rate under changing temperature. In the summer of 1993 (May 28 - August 30), we measured the air temperature at 1-hour intervals, and the surface snowmelt, in order to calculate a 'Degree Day Factor'. The surface snow melt was measured as the exposed length of a bamboo ping, inserted into the snow in a measured period. The snow density was about 0.6, almost stable during the observed period.

Figure 3 shows the relationship between the surface snowmelt (in snowdepth) and the warm index (WI). The WI is the accumulated excess of daily mean air temperatures to -3°C (Kojima, 1983). Therefore WI is zero when the daily mean temperature is -3°C. The length of each measured periods is from 3 to 9 days, and WI and surface snowmelt (M) is plotted for each period. The relationship between the total surface snowmelt (7,545mm) and the total WI (784.7 degree-day) in the observed period, is accurately represented by the following linear equation,

$$M = 9.6\ WI$$

where M is surface snow melt (mm; in snow depth), and the Degree Day Factor was 9.6 (mm/degree-day). Figure 4 shows the cumulative surface snowmelt curves as calculated from the above equation, and the mean air temperature of the last 16 years (1979-1994), as estimated from the meteorological data of Tazawako, near Mt. Zarumori (about 12 km southwestward). The closed circle indicates the innermost buried peat sites (snow release in early August). If the snowdepth was the same as

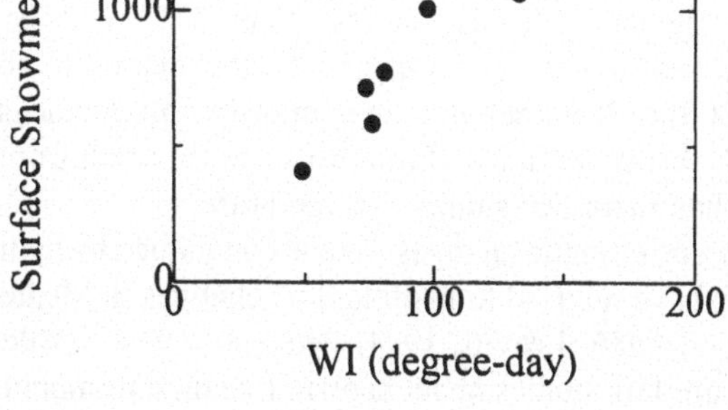

Fig. 3. Relationship between Warm Index (WI) and surface snowmelt depth.

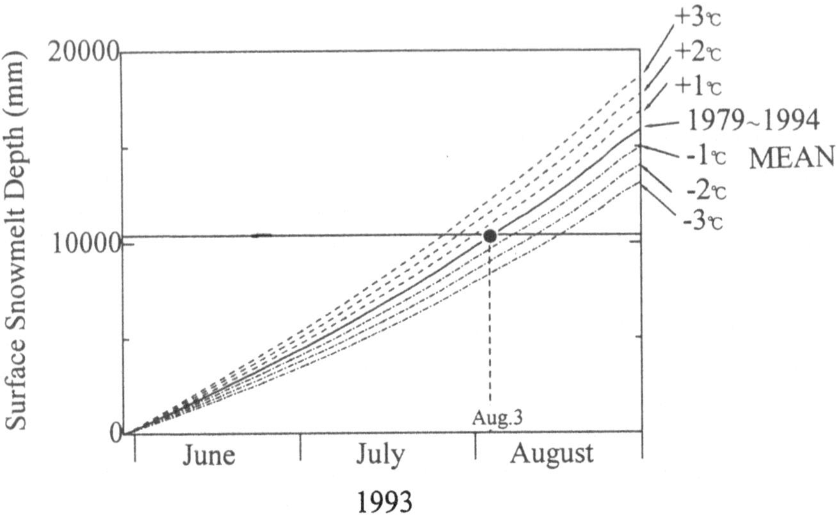

Fig. 4. Simulation of snowmelt curve caused by changes in summer air temperature.

1993, and the summer air temperature was 3° C warmer than during the last 16 years, then the snow cover will disappear by July 22. But this is still late for peat deposition, because at the latest, snow disappearance by the middle of July is required. This indicates that a 3° C warming will still be insufficient for the formation of buried peat, if the snow depths remains unchanged. But a 3° C warming will be an overestimation, because the Medieval era's warming is considered a relatively modest change (Hughes, 1994; Lamb, 1977; Yoshino, 1978). De'er (1994) estimated that the annual temperature during Medieval Time in South China was almost 1° C higher than today.

Figure 5 shows the shape of the snowbank in the studied site, during the summer of 1993 and 1994. The positions of the breakline are very different from year to year. Accordingly, if the amount of snow fall decreases, then the position of the break line will retreat and the snowdepth around the breakline will change remarkably for several meters. If the Medieval Period's warming was accompanied by decreased snowfall, a smaller temperature rise of less than 3° C, will be enough for the deposition of buried peat to occur. Thus, the overestimation in above 'Static Snow Model' will suggest the decrease of winter snowfall in Medieval Warm Period. Summer snow patch shrinkage in Mt. Zarumori seems to result from rather decreased snowfall in winter than increased summer temperature.

In general, Holocene climatic changes were accompanied by hydrological changes. A number of studies have referred to hydrological changes in Medieval Warm Period (Chatters and Hoover, 1986; Leavitt, 1994; Peterson, 1994; Stahle and Cleaveland, 1994). But, there are few studies about snowfall change in mountain area, because of limited proxies. Some studies presented that little ice age snow avalanche activity, good indicator of past snowfall, was higher than present day (Grove, 1972; Matthews

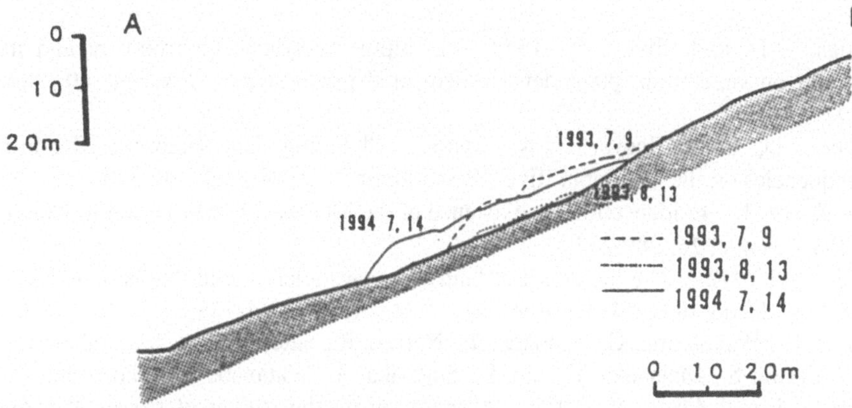

Fig. 5. The profiles of remnant snowbank, summer of 1993 and 1994.

and McCarroll, 1994).

In Japanese mountains, Sakaguchi (1982) has suggested a warming in Medieval Time by palynological study of subalpine peatland. But the studies about snowfall change during Holocene in mountain area are few. Yoshida *et al.* (1990) reported fossil ice dated 1,000-1,700 y.B.P. in Mt. Tateyama, Northern Japan Alps. Koizumi *et al.* (1983) reported the buried peat layer of neoglacial age (ca. 3,100 y.B.P.) formed by elongated snowmelt water. These study might suggest the increase of snowfall in colder epoch of Holocene. But snowfall in warmer epoch of Holocene time was not well known. Our study showed that the meadow soil stratigraphy in nivation hollows is very sensitive to so slight climatic changes as the Medieval's warming. With understanding of snow accumulation and snowmelt process of snow patches, we can discuss about the change of past snowfall by meadow soil stratigraphy. Also, it will be useful to predict future vegetational change, because alpine-subalpine vegetation in Japan is strongly affected by snow distribution.

5. CONCLUSION

The buried peats found in the subalpine meadow soil of Mt. Zarumori, indicate that the summer was rather warm and that the winter had decreased snowfall during the Medieval time.

If the probable warming by an intensification of the greenhouse effect exceeds 2℃, snow patches and alpine meadows in Japanese subalpine zones will shrink considerably. If the warming is accompanied by decreased snowfall and snow accumulation, the shrinkage will be accelerated and many meadows will disappear.

REFERENCES

Billings, W.D. and Bliss, L.C., 1959. An alpine snowbank environment and its effects on vegetation, plant development, and productivity. *Ecology,* **40**, 388-397.

Chatters, J.C. and Hoover, K.A., 1986. Changing late Holocene flooding frequencies on the Columbia River, Washington. *Q. Res.*, **26**, 309-320.

De'er, Z., 1994. Evidence for the existence of the Medieval Warm Period in China. *Clim. Change*, **26**, 289-297.

Grove, J.M., 1972. The incidence of landslides, avalanches, and floods in Western Norway during the Little Ice Age. *Arct. Alp. Res.*, **4**, 131-138.

Higuchi, K., Wakahama, G., Yamada, T., Naruse, R., Sato, S., Abe, M., Nakayama, S., Koiwa, S., Matsuoka, H., Ito, F., Sagisaka, S., Watanabe, O., Nakajima, C., Inoue, J. and Ageta, Y., 1979. A summary on the studies of perennial snow patches in Japan. *J. Jpn. Soc. Snow Ice (Seppyo)*, **41**, 181-197. (in Japanese with English abstract).

Hughes, M.K., 1994. Was there a 'Medieval Warm Period', and if so, where and when. *Clim. Change*, **26**, 109-142.

Japan Association for Quaternary Research, 1987. *Quaternary Maps of Japan.* Univ. Tokyo Press, Tokyo. (in Japanese with English legend).

Kariya, Y., 1994. Genesis of buried muck in the subalpine zone of Mount Gassan, snowy mountain in northern Japan. *Daiyonki-Kenkyu* (*Q. Res.*), **33**, 269-276 (in Japanese with English summary).

Koizumi, T., 1982. Increase in snowfall since late Glacial Age in the mountainous region of Japan, viewed from fossil periglacial slopes, nivation hollows and mountain oligotrophic bogs. *Daiyonki-Kenkyu* (*Q. Res.*), **21**, 245-253. (in Japanese with English summary).

Koizumi, T., Yamanaka, N., Hara, A. and Sakamoto, S., 1984. Climatic change in the late Holocene inferred from the buried peat layer in Mt. Tairappyo, Mikuni range, central Japan. *Geogr. Rev. Jpn.*, **57** (Ser.A), 739-748. (in Japanese with English summary).

Kojima, K., 1983. Estimation of melting rate of snow by simple formulate using only air temperature. *Low Temp. Sci.*, **42** (Ser.A), 101-110. (in Japanese with English summary).

Lamb, H.H., 1977. *Climate: Present, Past and Future. Vol. 2, Climatic History and the Future.* Methuen & Co., London, 835 pp.

Leavitt, S.W., 1994. Major wet interval in White Mountains medieval warm period evidenced in $\delta^{13}C$ of bristlecone pine tree rings. *Clim. Change*, **26**, 299-307.

Matthews, J.A. and McCarroll, D., 1994. Snow-avalanche impact landforms in Breheimen, Southern Norway: Origin, age, and paleoclimatic implications. *Arct. Alp. Res.*, **26**, 103-115.

Mizota, C. and Matsuhisa, Y., 1985. Eolian additions to soils and sediments of Japan. *Soil Sci. Plant Nutr.*, **31**, 369-382.

Morioka Local Meteorological Observatory, 1966. *Climatography of Iwate Prefecture.*

268 pp.

Osumi, Y. and Kumada, K., 1971. Studies on alpine soils in Japanese North Alps (part 4) ^{14}C-Dating of the Muck-like horizon of alpine wet meadow soils and the buried soils of alpine grassland soils. *J. Sci. Soil Manure Jpn.*, **42**, 270-272. (in Japanese).

Peterson, K.L., 1994. A warm and wet Little Climatic Optimum and a cold and dry Little Ice Age in the Southern Rocky Mountains, U.S.A. *Clim. Change*, **26**, 243-269.

Sakaguchi, Y., 1982. Climatic variability during the Holocene epoch in Japan and its causes. *Bull. Dep. Geogr. Univ. Tokyo*, **14**, 1-27.

Stahle, D.W. and Cleaveland, M.K., 1994. Tree-ring reconstructed rainfall over the Southern U.S.A. during the medieval warm period and little ice age. *Clim. Change*, **26,** 199-212.

Stanton, M.L., Rejma'nek, M. and Galen, C., 1994. Changes in vegetation and soil fertility along a predictable snowmelt gradient in the Mosquite Range, Colorado, USA. *Arct. Alp. Res.*, **26**, 364-374.

Watson, A., Davison, R.W. and French, D.D., 1994. Summer snow patches and climate in Northeast Scotland, U.K. *Arct. Alp. Res.*, **26**, 141-151.

Yamanaka, H., 1983. Radiocarbon ages of alpine wet-meadow soil on Mt. Iide. *Daiyonki-kenkyu* (*Q. Res.)*, **21**, 315-321. (in Japanese).

Yoshida, M., Yamamoto, K., Higuchi, K., Iida, H., Ohata, T. and Nakamura, T., 1990. First discovery of fossil ice of 1000-1700 year B.P. in Japan. *J. Glaciol.*, **36**, 258-259.

Yoshino, M., 1978. Regionality of climatic change in monsoon Asia: In: *Climatic Change and Food Production* (ed by M.M. Yoshino and K. Takahashi), Univ. Tokyo Press, Tokyo, 331-341.

EFFECTS OF EXTERNAL FACTORS ON GROWTH AND STRUCTURE OF *Pinus pumila* SCRUB IN MT. KINPU, CENTRAL JAPAN

Takuya Kajimoto[1], Nahoko Kurachi[2], Yukihiro Chiba[3], Hajime Utsugi[3] and Moriyoshi Ishizuka[3]

ABSTRACT: Site differences in microclimates and productivity due to slope aspect and elevation were assessed for the alpine dwarf pine (*Pinus pumila* Regel) scrub on Mt. Kinpu, central Japan. Leaf area index, aboveground biomass and annual production rates were larger for the pine stands on the south-facing slope than those on the north-facing slope. For each slope, these parameters increased with decreasing elevation. With respect to environmental factors, the daily photon flux density during the summer months was almost the same between the N- and S-facing slopes, but the temperature conditions differed due to slope aspect, especially soil-temperature, which was 2-4°C higher on the S-facing slope. The positive correlation between temperature and productivity indicated that stand development of *P. pumila* scrub could be enhanced under higher temperatures. The effects of recent global warming on the Japanese alpine/subalpine vegetation dynamics were discussed by focusing on habitat competition between *P. pumila* and other subalpine conifers.
KEY WORS: *Pinus pumila* scrub, microclimate, productivity, subalpine conifers.

1. INTRODUCTION

Among environmental factors, air- and soil-temperatures are what primarily affect growth and photosynthetic production of conifer trees growing in high elevations (e.g., Turner *et al.*, 1983; Häsler, 1985; Delucia and Smith, 1987). It has often been suggested that the trees around the alpine timberline are susceptible to desiccation damage because of incompletely developed cuticles resulting from the combined effects of low temperature and short growing period, the main factors causing the elevational tree-lines (Wardle, 1971; Tranquillini, 1979; Delucia and Berlyn, 1984). The regeneration process of subalpine conifers, e.g., cone production, seed dispersal and seedling establishment, may also depend on summer

[1] Tohoku Research Center, Forestry and Forest Products Research Institute, 72 Nabeyashiki, Shimokuriyagawa, Morioka, Iwate 020–01, Japan.
[2] Hokkaido Research Center, Forestry and Forest Products Research Institute, Hitsujigaoka–7, Toyohira–ku, Sapporo, Hokkaido 062, Japan.
[3] Forestry and Forest Products Research Institute, P.O. Box 16,Tsukuba Norin, Ibaraki 305, Japan.

temperature conditions. The upward shifts of the tree-limits reported for some spruce trees in subarctic regions were caused by extensive establishment and growth of seedlings in warmer years (Payette and Filion, 1985; Kullman, 1986). Thus, the elevational distribution of subalpine conifer species might be governed by temperature conditions.

In central Japan, the uppermost areas of high mountains are generally covered with a dwarf scrub of *Pinus pumila* Regel, below which are extensive subalpine conifer forests consisting of *Abies veitchii*, *A. mariesii* and *Tsuga diversifolia*. Due to global warming, remarkable in recent decades (e.g., Hansen *et al.*, 1981; Jones *et al.*, 1986), these subalpine conifers are likely to migrate upward into the *P. pumila* zone. However, the temperature rise may also affect the population dynamics of the pine scrub. To predict future vegetational changes in the Japanese alpine/subalpine regions, information on the environmental conditions as they affect the growth, survival and regeneration process of each species should be collected.

This paper assesses the relationships between external factors (irradiance and temperature) and productivity of *P. pumila* scrub based on the local site differences in these parameters. It is discussed future vegetational changes caused by global warming in the *P. pumila* zone and in the subalpine conifer forests.

2. MATERIALS AND METHODS

2.1 Study site

The study was carried out on Mt. Kinpu (2,595 m a.s.l.) in the Okuchichibu mountain range, central Japan (Fig. 1). The uppermost part of Mt. Kinpu (above ca. 2,450 m) was dominated by a dwarf scrub of *P. pumila*, below which (above ca. 1,700 m) was represented by subalpine conifer forests consisting mainly of *Abies mariesii*, *A. veitchii* and *Tsuga diversifolia*. Six research plots (1-2 m^2 in area) were established in the pine scrub; three plots were on the north-facing slope at different elevations (N1: 2,575 m, N2: 2,560 m, N3: 2,540 m) and the other three plots were on the south-facing slope (S1: 2,580 m, S2: 2,560 m, S3: 2,550 m).

2.2 Field measurements

The stand structures and productivities of the six plots were assessed by field census based on a unit of aboveground stem of *P. pumila* (Kajimoto, 1989). All *P. pumila* stems in each plot were measured for their basal diameters (D), diameters at the lowest living branch (D_B), stem lengths (L) and heights of terminal leader shoots. Eleven stems of different sizes (n=6 in the north-facing plots, n=5 in the south-facing plots) were selected for a detailed stem census. All leafy shoots of each sample stem were classified into three shoot-groups; terminal leader of stem, terminal leader of first-order branches and other lateral shoots. The shoots were measured for their aboveground heights, shoot lengths and the number of fascicles with living needles. The shoot length was measured separately by age, i.e., the age of shoot was decided by counting the number of nodes corresponded to the yearly growth. The needle length of each shoot was also determined by age as the average of five fascicles (5 needles in a facicle). The field censuses were conducted in the

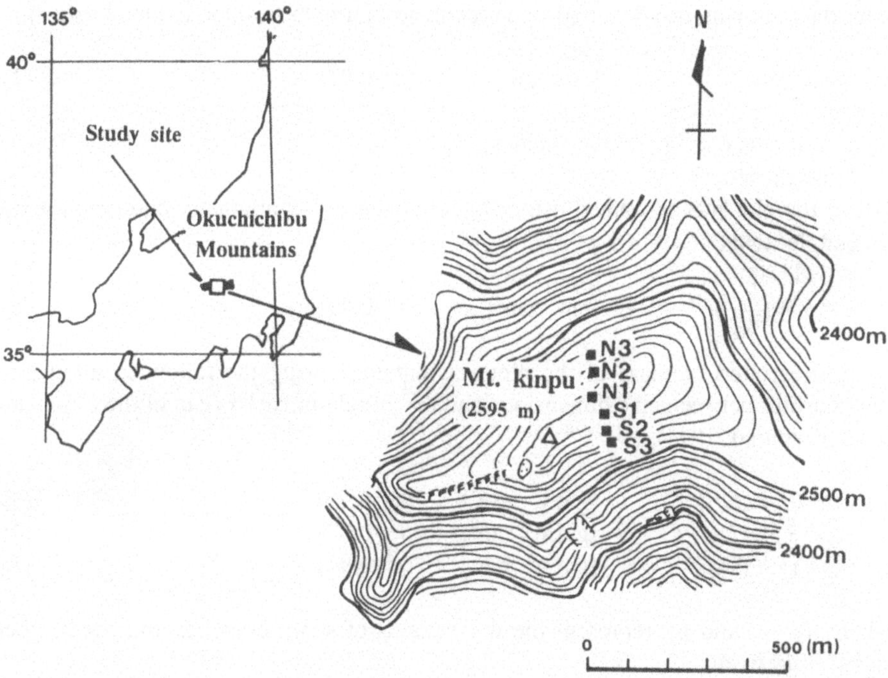

Fig. 1. Location of study site and arrangement of six research plots of *Pinus pumila* scrub at Mt. Kinpu.

summers of 1992 and 1993.

To determine the attenuation process of solar irradiance within the canopy, the relative value of photon flux density at different heights (5 cm intervals) was measured in each plot with two quantum sensors (LI-190SA, Licor, USA) connected to a data logger (LI-1000, Licor).

Environmental conditions were monitored at the highest plots of each slope aspect (N1 and S1) from June to early October in 1992. Photon flux density above the canopy surface was measured with quantum sensors (IKS-25, Koito, Japan) and data loggers (UP, Kadec, Japan) at 10-minute intervals. Inside air-temperature (30 cm in height) and soil-temperature (10 cm in depth) were recorded with thermister sensors (KDC-S1, Kadec) and data loggers (UV, Kadec) at 30-minute intervals.

2.3 Estimation of leaf area index and biomass

In this study, leaf area index (*LAI*) and aboveground biomass of the research plots were estimated using the field census data mentioned above and the destructive sampling data of *P. pumila* stems and needles that had been obtained at other research areas (Mt. Shougigashira, Kiso Mountains) (Kajimoto, 1989, 1994).

Total leaf areas (u) (projected area) of the 11 sample stems of the detailed census were obtained as the sum of leaf area in each shoot. Leaf area of each shoot was estimated by multiplying the number of fascicles by mean leaf area that was obtained from the following needle length-area regression formula (Kajimoto, unpublished):

$$s = 0.23 \cdot l^{1.40} \quad [\mathrm{cm}^2, \mathrm{cm}] \quad (r^2=0.97) \tag{1}$$

s : leaf area of a fascicle (5 needles)
l : needle length.

LAI of the plot was estimated using the following D_B^2-u allometry observed for the 11 sample stems,

$$u = 554 \cdot (D_B^2)^{0.93} \quad [\mathrm{cm}^2, \mathrm{cm}^2] \quad (r^2=0.88). \tag{2}$$

Aboveground biomass of the plot was obtained using the following allometric relationships between size dimensions and dry weight of each organ of the *P. pumila* stem (Kajimoto, 1994),

$$w_S = 0.51 \cdot (D^2 \cdot L)^{0.88} \quad [\mathrm{g\ d.w.}, \mathrm{cm}^3] \quad (r^2=0.99) \tag{3}$$
$$w_B = 5.38 \cdot (D_B^2)^{1.56} \quad [\mathrm{g\ d.w.}, \mathrm{cm}^2] \quad (r^2=0.87) \tag{4}$$
$$w_L = 17.6 \cdot (D_B^2)^{1.02} \quad [\mathrm{g\ d.w.}, \mathrm{cm}^2] \quad (r^2=0.95) \tag{5}$$

where w_S, w_B and w_L represent the dry weights of stem, branches and needles per stem, respectively.

3. RESULTS

3.1 Stand structure

Among the six research plots, the size of aboveground stems tended to increase with decreasing stem density (Table 1). Annual mean elongation rate of terminal shoots ranged from 1.4 to 2.0 cm y^{-1}. Stand ages of the six plots, defined as the ratio of mean stem length to elongation growth rate, were estimated to be 26-55 yrs.

Estimates of aboveground total biomass in the south-facing three plots reached 27-40 ton ha^{-1} somewhat larger than the 12-33 ton ha^{-1} in the north-facing three plots (Table 1). *LAI* values of the south-facing plots (2.4-4.0 m^2 m^{-2}) were also larger than those of the north-facing plots (1.7-2.6 m^2 m^{-2}). For each slope, especially the south-facing slope, the biomass and *LAI* increased as elevation decreased.

Relative values of photon flux density (*PFD*) just below the canopy layers were 31% (N1), 39% (N2), 18% (N3) in the plots of north-facing slope and 25% (S1), 19% (S2), 15% (S3) in the south-facing slope. The difference in relative *PFD* values was related to *LAI* values of the plots, i.e., the *PFD* decrease with increasing *LAI*.

3.2 Environmental conditions and productivity

Table 2 shows environmental conditions observed at the highest plots on the north-

Table 1 Results of stem census and estimates of aboveground biomass and leaf area index (*LAI*) in the *Pinus pumila* stands at Mt. Kinpu

	N-facing slope			S-facing slope		
Plot	N1	N2	N3	S1	S2	S3
Elevation (m)	2,575	2,560	2,540	2,80	2,560	2,550
Scrub height (cm)	40	30	70	50	40	50
Stem density (m^{-2})	21	33	14	13	32	23
Mean basal diameter (cm)	1.5	1.0	2.3	2.1	1.5	2.0
Mean stem length (cm)	58	37	83	74	51	57
Elongation growth rate (cm y^{-1})*	1.5	1.4	1.5	2.0	1.4	1.5
Stand age (yrs-old)**	39	26	55	37	36	38
Biomass (ton ha^{-1})						
Stems and Branches	12.7	6.2	22.4	17.1	18.3	23.7
Needles	7.8	5.6	10.4	9.7	12.6	15.8
Aboveground total	20.5	11.8	32.8	26.8	30.9	39.5
LAI (m^2 m^{-2})	2.1	1.7	2.6	2.4	3.5	4.0

*Mean annual shoot length of terminal leader during the recent 3 yrs. (1990-1992) averaged for the stems in each plot.

**Mean age of aboveground stems estimated by the ratio of mean stem length to elongation growth rates.

facing (N1) and south-facing (S1) slopes. Daily mean values of *PFD* above the canopy surface in the summer months (July, August and September) were almost the same between N1 and S1. Total *PFD* during the growing season from June to early October (127 days) was 3799 mol m^{-2} in S1 that was only 3% larger than 3681 mol m^{-2} in N1. Mean air temperatures in the three summer months were 1-2°C higher on the south-facing slope (S1) than on the north-facing slope (N1) (Table 2). Soil temperatures were 2-4°C higher in S1 than in N1. The warmth index of each plot, which was calculated from the monthly mean air temperatures above 5°C, reached ca. 20 °C month.

For the stand productivity of *P. pumila*, two indices were determined, i.e., annual mean biomass growth rates of woody organs and annual needle production. The biomass growth rates, defined as the ratio of woody biomass (stems and branches) to stand age (Table 1), were about two times larger in the south-facing three plots (0.5-0.6 ton ha m^{-1} y^{-1}) than in the north-facing plots (0.2-0.4 ton ha^{-1} y^{-1}) (Table 2). Needle production, which was assumed to be equal to the current needle biomass, was slightly larger in the south-facing plots (2.4-4.0 ton ha^{-1} y^{-1}) than in north-facing plots (1.9-3.5 ton ha^{-1} y^{-1}).

4. DISCUSSION

In the European Alps, effects of microclimates on the photosynthesis of subalpine conifer trees have been well examined on two contrasting slopes (e.g., north- and east-facing) of the same elevation (Tranquillini, 1979). Turner *et al.* (1983)

Table 2 Environmental conditions and estimates of productivity in the *P. pumila* stands at Mt. Kinpu. Photon flux density (*PFD*) and temperatures were measured at the two plots, N1 (N-facing slope) and S1 (S-facing slope) in 1992. Values of productivity were the ranges for the three plots on each slope.

		N-facing slope	S-facing slope
Daily mean *PFD* (mol m^{-2} d^{-1})	July	36.8	37.7
	Aug.	27.2	27.7
	Sept.	26.3	28.6
Mean air-temperature (°C)	July	12.4	13.8
	Aug.	12.5	13.1
	Sept.	7.8	9.9
Mean soil-temperature (°C)	July	9.6	12.3
	Aug.	11.6	13.2
	Sept.	7.4	10.0
Warmth index (°C month)		20.2	21.7
Productivity (ton ha^{-1} y^{-1})			
Biomass increment rate*		0.2 - 0.4	0.5 - 0.6
Needle production rate**		1.9 - 3.5	2.4 - 4.0

* Defined as the ratio of aboveground woody biomass (stems and branches) to the stand age.
**Defined as the current needle biomass; the ratio of needle biomass to mean needle longevity. Needle longevities were 3 yrs in the N-slope and 4 yrs in the S-slope (Kajimoto, unpublished data).

indicated that the annual net carbon gains of two pine trees (*Pinus montana, P. cembra*) were much less on the north-facing slope than on the east-facing slope due to the limitations of soil-temperature and total photosynthesis period on the north-facing slope. Häsler (1985) also showed that the average daily CO_2 uptake rates of *Larix decidua* seedlings on the north-facing slope were about half of those on the east-facing slope irrespective of similar irradiance input, and suggested that the lower soil temperature level on the north-facing slope was the main limiting factor in photosynthesis. Similarly, the productivity of *P. pumila* stands was greater on the south-facing slope than on the north-facing slope (Table 2). This might be due in large part to the higher summer temperatures on the south-facing slope, especially soil-temperature, because the differences in microclimates due to slope aspect were more conspicuous in temperature than in irradiance regimes (Table 2).

In the subarctic regions, the upward shifts of tree-limits due to extensive seedling establishment in the warmer years of the 20th century have been observed for *Picea glauca* (Payette and Filion, 1985) and *P. abies* (Kullman, 1986). Temperature increase caused by recent global warming, which was noted in the 1980s (e.g., Hansen *et al.*, 1981; Jones *et al.*, 1986), may promote the migration of Japanese subalpine conifers (*A. veitchii, A. mariesii* and *Tsuga diversifolia*) into higher elevations. However, the positive correlation between temperature and productivity mentioned above indicates that stand development of the *P. pumila* scrub could be enhanced under higher temperatures. Thus, in the Japanese alpine regions, regeneration success of subalpine conifers primarily depends on habitat competition with *P. pumila*.

Kimura (1963) showed that the relative light intensities on the floors of subalpine fir (*A. veitchii*, *A. mariesii*) and hemlock (*T. diversifolia*) forests averaged 5.9-6.5% allowing the survival of seedlings of each tree species. Kohyama (1983) suggested that *A. veitchii* and *A. mariesii* seedlings required a relative light intensity of 5-8% for survival and 27-28% for maintaining normal growth-rates to reach the sapling stage. Generally, the natural regeneration of the *Abies* and *Tsuga* can progress only in open sites after gap-formation. These results in relatively even-aged forests, such as "wave-regenerated *Abies* forests" (Kohyama, 1982) and "mosaic-patch *Tsuga* forests" (Kanzaki, 1984). In the *P. pumila* stands at Mt. Kinpu, light conditions under the canopies, 15-39% in relative *PFD*, may allow the establishment of such conifer seedlings.

Kajimoto (1989) showed that mature *P. pumila* stands (ca. 100 yrs-old in the aboveground age) in the Kiso Mountains had *LAI* values exceeding 5 $m^2\ m^{-2}$, in which the relative *PFD* values on the floors were below 10%. Kajimoto (1992) also indicated that gap-formation due to individual mortality had rarely occurred within the mature *P. pumila* scrub because of its vegetative propagation by layering. Such canopy structure and reproduction habits in the well-developed pine scrub probably act defensively against invasion from other subalpine conifers (Kajimoto, 1994). Thus, it is impossible to predict future vegetational changes in the Japanese alpine/subalpine areas, i.e., upward expansion of the subalpine conifer forests to succeed the *P. pumila* zone, based solely on temperature increases resulting from global warming.

5. CONCLUSION

This study discussed the effects of temperature increase caused by global warming, especially on the population dynamics of *P. pumila* scrub in relation to the predicted upward shift of subalpine conifer trees. It suggested that subalpine conifers expand into alpine regions, but hardly succeed the vegetation zone where well-developed *P. pumila* scrub had already occupied, even under the higher temperature conditions. However, the increase in air- and soil-temperatures might be linked with changes in other factors, e.g., precipitation and soil conditions (water, nutrients). Information on such environmental factors affecting growth and regeneration for both *P. pumila* and the subalpine conifers will be required for further discussion of vegetation changes in the Japanese alpine/subalpine zones.

REFERENCES

Delucia, E.D. and Berlyn, G.P., 1984. The effect of increasing elevation on leaf cuticle thickness and cuticular transpiration in balsam fir. *Can. J. Bot.*, **62**,

2423-2431.

Delucia, E.D. and Smith, W.K., 1987. Air and soil temperature limitations on photosynthesis in Engelmann spruce during summer. *Can. J. Forest Res.*, **17**, 527-533.

Hansen, J., Jhonson, D., Lacis, A., Lebedeff, S., Lee, P., Rind D. and Russell, G., 1981. Climate impact of increasing atmospheric carbon dioxide. *Science*, **213**, 957-966.

Häsler, R., 1985. CO_2 uptake in European larch (*Larix decidua* Mill.) near treeline in Switzerland (Stillberg/Davos). In: *Establishment and Tending of Subalpine Forest: Research and Management* (ed by H. Turner and W. Tranquillini), Proc. IUFRO Workshop 1984, Eidg. Anst. Forstl. Versuchswes., Ber. **270**, 113-122.

Jones, P.D., Wigley, T.M.L. and Wright., P.B., 1986. Global temperature variations between 1861 and 1984. *Nature*, **322**, 430-434.

Kajimoto, T., 1989. Aboveground biomass and litterfall of *Pinus pumila* scrubs growing on the Kiso mountain range in central Japan. *Ecol. Res.*, **4**, 55-69.

Kajimoto, T., 1992. Dynamics and dry matter production of *Pinus pumila* trees growing on the Kiso mountain range in central Japan. *Ecol. Res.*, **7**, 333-339.

Kajimoto, T., 1994. Aboveground net production and dry matter allocation of *Pinus pumila* forests in the Kiso mountain range, central Japan. *Ecol. Res.*, **9**, 193-204.

Kanzaki, M., 1984. Regeneration in subalpine coniferous forests. 1. Mosaic structure and regeneration process in a *Tsuga diversifolia* forest. *Bot. Mag. Tokyo*, **97**, 297-311.

Kimura, M., 1963. Dynamics of vegetation in relation to soil development in north-facing Yatsugatake mountains. *Jpn. J. Bot.*, **18**, 255-287.

Kohyama, T., 1982. Studies on the *Abies* population of Mt. Shimagare. 2. Reproductive and life history traits. *Bot. Mag. Tokyo*, **95**, 167-181.

Kohyama, T., 1983. Seedling stage of two subalpine *Abies* species in distinction from sapling stage: A matter-economic analysis. *Bot. Mag. Tokyo*, **96**, 49-65.

Kullman, L., 1986. Recent tree-limit history of *Picea abies* in the southern Swedish Scandes. *Can. J. Forest Res.*, **16**, 761-771.

Payette, S. and Filion, L., 1985. White spruce expansion at the tree line and recent climatic change. *Can. J. Forest Res.*, **15**, 241-251.

Tranquillini, W., 1979. *Physiological Ecology of the Alpine Timberline*. Springer-Verlag, Berlin, 131 pp.

Turner, H., Häsler, R. and Schönenberger, W., 1983. Contrasting micro-environments and their effects on carbon uptake and allocation by young conifers near alpine treeline in Switzerland. In: *Carbon Uptake and Allocation in Subalpine Ecosystem as a Key to Management* (ed. by R.H. Wareing), Proc. IUFRO Workshop 1982, Oregon State University, Forestry Research Laboratory, 22-30.

Wardle, P., 1971. An explanation for alpine timberline. *N. Z. J. Bot.*, **9**, 371-402.

EFFECTS OF ENVIRONMENTAL CONDITIONS ON THE SITE DYNAMICS OF A SUBALPINE FOREST IN THE AKAISHI MOUNTAINS, JAPAN

Shigeo Kobayashi[1], Hisao Sakai[1], Nagaharu Tanaka[1] and Kiyoshi Fujimoto[1]

ABSTRACT: This paper discusses probable effects of global warming on the site dynamics of a subalpine forest from a point of view that changes in soil environments affect the distribution of vegetation established their community on the peculiar soil. The study area is located in the Akaishi Mountains from Sawarajima to Fujimitaira in Shizuoka Prefecture. Data were collected and measured on microclimates, stand structure and growth, composition of understory vegetation, litter decomposition rate, soil properties and tree regeneration at seven fixed plots. Air temperature in November decreased linearly with increasing elevation because of adiabatic expansion of air with its ascent. Soil temperature showed a clear decrease when elevation increased, except in the highest plot (el.2725 m). The pH values in the topsoil indicated a negative correlation with elevation in which the pH values decreased as elevation increased. The total carbon content in the topsoil tended to increase with elevation, except in the highest plot. The relationship between elevation and the decomposition rate did not show a clear negative correlation due to that it is affected by landform and vegetation.
KEY WORDS: subalpine, temperature, soil properties, litter decomposition.

1. INTRODUCTION

Global warming affects not only air temperature, but also the water regime in a forest ecosystem (Schlesinger *et al.*, 1990; EPA *et al.*, 1990). Nevertheless, the forest ecosystem is so complicated, it is difficult to verify that the increase in air temperature directly induces changes in vegetation distribution and increments in tree growth. Increasing air temperature and desiccation directly affect the amount of snowfall and the time of snow melt. It is also expected that the increase in temperature and desiccation affect the decomposition rate of litter accumulation and soil humus, and change the soil properties. It is also believed that changes in soil properties affect the distribution of plant communities. The relationship between *Tsuga diversifolia* forest and podzolic soils is typical of areas higher than 2,200 m above the sea level in the Akaishi Mountains (Kobayashi *et al.*, 1987). The

[1] Forestry and Forest Products Research Institute, P.O. Box 16, Tsukuba Norin, Ibaraki 305, Japan.

warming effects will be examined effectively only short term effect on the subalpine forest ecosystem.

Therefore, it is important to clarify probable effects of increasing air temperature and changes in water regime on the site conditions and the ecological niche of subalpine forests. The purpose of this study is to (1) clarify the site conditions including vegetation at different elevations, (2) clarify the microclimate such as air and soil temperature, precipitation and soil water conditions, (3) clarify the decomposition rate of *T. diversifolia* litter and related site conditions and microclimate, and (4) predict changes in site conditions to be caused by climatic change.

2. SITE DESCRIPTION AND METHODS

The study area is located in the Akaishi Mountains from Sawarajima to Fujimitaira in Shizuoka Prefecture. Its elevation ranges between 1,120 m (Sawarajima) and 2,725 m (Fujimitaira) above sea level (Fig. 1). The Akaishi Mountains mainly consist of Mesozoic sedimentary rocks which were modified by Shimanto activity. Temperature varies depending on topography and elevation with an annual mean of 4 - 5 at 2000 m above sea level. Annual precipitation is about 2,600 mm at the same elevation.

Seven fixed plots were set at each elevation, i.e., plot No. 1 at 1,120 m, plot No. 2 at 1,560 m, plot No. 3 at 1840 m, plot No. 4 at 2,130 m, plot No. 5 at 2,470 m, plot No. 6 at 2560 m and plot No. 7 at 2,725 m (Table 1). Landforms are classified into five types, i.e., alluvial toeslope, convex creepslope, transportational midslope, seepageslope and interfluve (Dalrymple *et al.*, 1968; Conacher and Dalrymple, 1977). Soil types which appeared in these experimental plots are B_D, B_C (P_DIII), B_A (P_DIII), P_DII, P_DI (Forest Soil Division, 1976). Forest types are classified into *Larix*

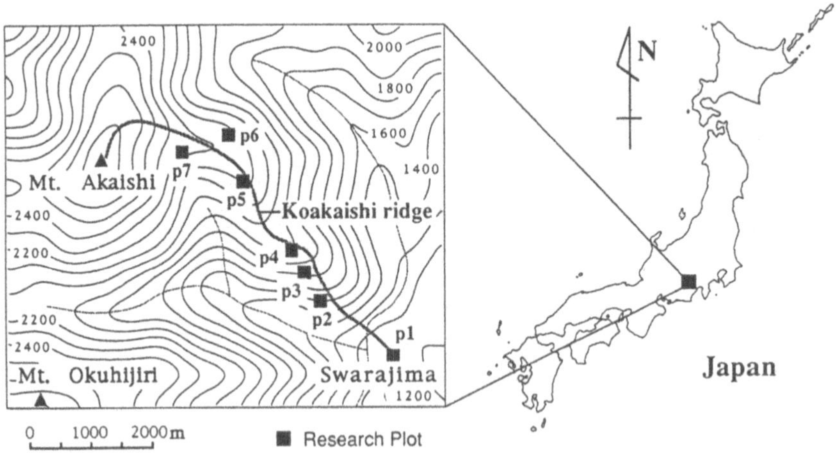

Fig. 1. Topography and location of experimental plots.

Table 1 Description of each experimental plot in the Akaishi Mountains

Plot No.	Elevation	Landform	Soil type	Forest type	Understory type
1(Sawarajima)	1120m	Alluvial toeslope	B_D	*Larix kaempferi*	Macro herbs
2	1560m	Convexcreepslope	Bc (P_DIII)	*Quercus crispula*	no indicator
3	1840m	Transportational midslope	B_A (P_DIII)	*Tsuga diversifolia*	no indicator
4	2130m	Seepage slope	B_A (P_DIII)	*Abies veittchii*	*Dryopteris-Cacalia*
5	2470m	Seepage slope	P_DII	*Abies veittchii*	Moss
6	2560m	Interfluve	P_DI (P_DII)	*Abies mariesii*	*Rhododendron*
7(Fujimitaira)	2725m	Interfluve	P_DI	*Pinus pumila*	*Rhododendron*

1. Monitoring of microclimate
   : Precipitation, soil water tension, air and soil temperature
2. Tree census
   : GBH, height, (seedling), understory vegetation type
3. Liter decomposition survey
   : Litter bag, (litter fall)
4. Soil survey
   : Soil profile, soil physical and chemical properties, A_0

kaempferi, *Quercus crispula*, *Tsuga diversifolia*, *Abies veittchii*, *Abies mariesii* and *Pinus pumila* with understory vegetational types of macro herbs, *Dryopteris-Cacalia*, moss, and *Rhododendron* (Maeda and Miyakawa, 1970; Kobayashi *et al.*, 1987).

Data were collected on microclimate (air temperature, soil temperature, soil water conditions and precipitation) using a Kadec data logger (KONA system), measured stand structure and growth, composition of understory vegetation, litter decomposition rate using litter bag, soil and tree regeneration at seven fixed quadrates. Litter bags were filled with 50 grams of twigs and leaves of *Tsuga diversifolia* and five bags were taken from every plot to determine the decomposition rate of litter.

3. RESULTS AND DISCUSSION

3.1 The relationship between climate conditions and elevation

Air temperature changed with elevation at the Koakaishi ridge in Akaishi Mountains, although plot No. 1 and plot No. 2 could not be measured (Fig. 2). In November, air temperature decreased linearly at the speed near the moist adiabatic lapse rate as elevation increased. Air temperature at plot No. 7 was higher than at plots No. 5 and No. 6 in May and June, and air temperature at plots No. 4 and No. 5 higher than at plot No. 3 in February and March. These temperature variations resulted from differences in microlandforms and vegetation. In addition, air temperature at plots No. 3 - No. 7 was maintained below 0 during the period December to March.

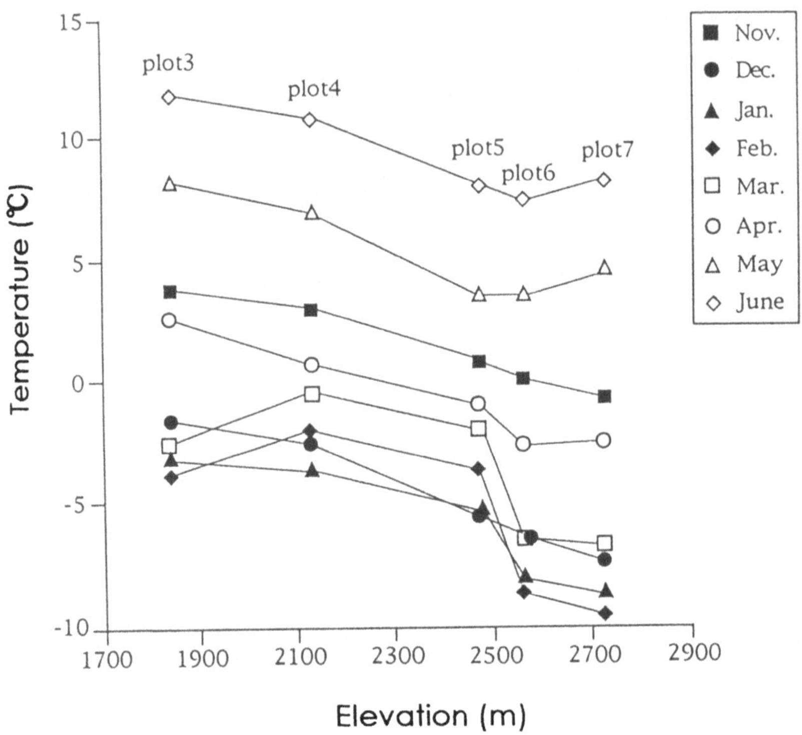

Fig. 2. Relation between elevation and air temperature.

Soil temperature showed also a clear decrease with increasing the elevation except at plot No. 7 (Fig. 3). Forest types in plots No. 1 to No. 6 contained taller tree species than at plot No. 7, where *Pinus pumila* was dominant. The taller and dense forest canopy intercepts well the penetration of sun light on the forest floor more than the lower and sparse forest canopy. Annual march of soil temperature in the subalpine forest showed typical patterns in which soil temperature starts to increase in May and decrease in October (Fig. 4). Soil temperature affects the decomposition of *Tsuga diversifolia* litter from May to October. From November to April, soil temperature falls below 0 and soil will be frozen, although the annual march of soil temperature fluctuates year by year.

These soil temperature conditions affect the litter decomposition rate and soil water regime.

3.2 Relationship between site conditions and elevation

Topsoil properties such as pH, total carbon, total nitrogen and C/N ratio are shown in Figs. 5, 6 and 7. pH values in the topsoil indicated a negative correlation with

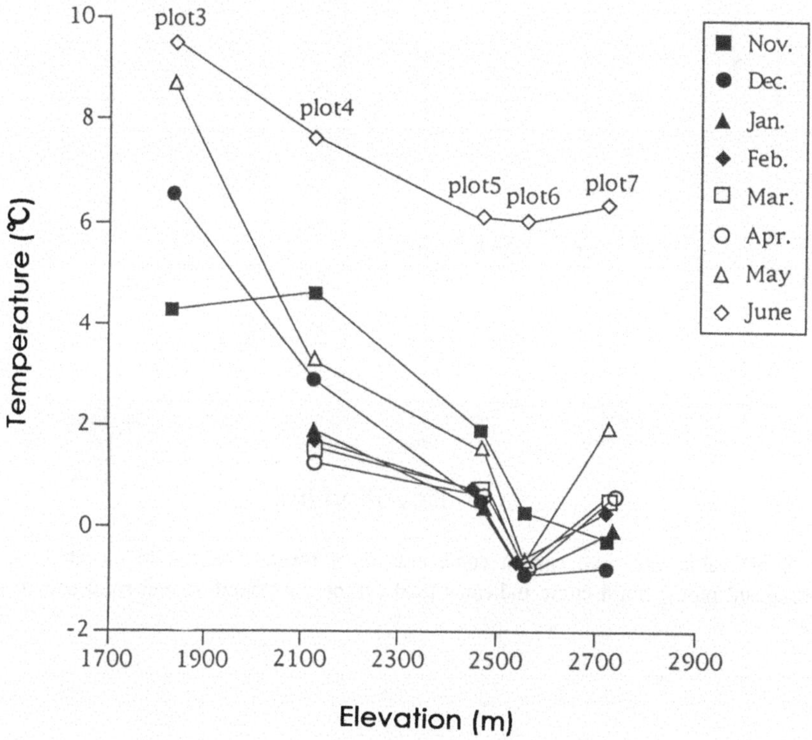

Fig. 3. Relation between elevation and soil temperature.

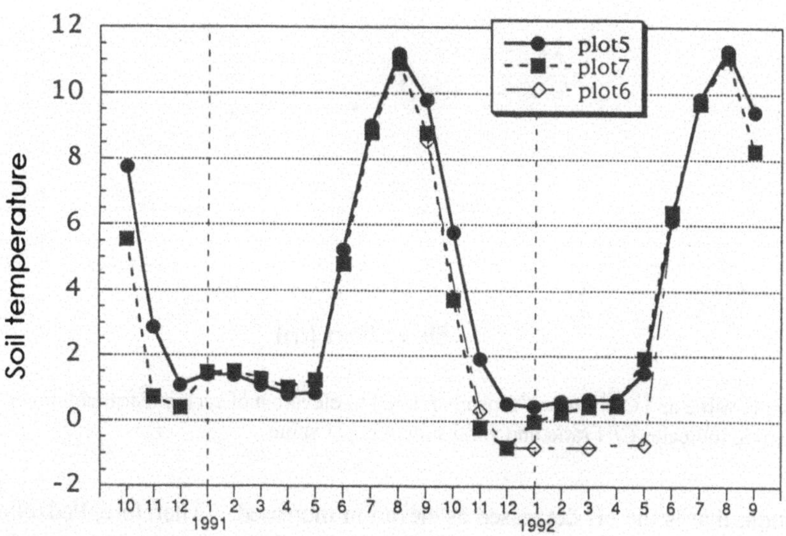

Fig. 4. Annual change in soil temperature at plot Nos. 5,6 and 7.

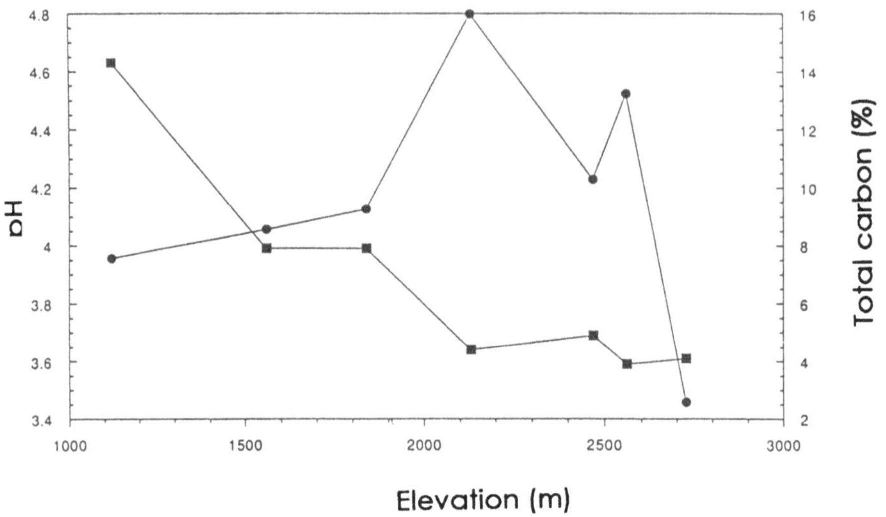

Fig. 5. pH value and total carbon concentration of topsoils related to elevation of each experimental plot. Solid circle indicates total carbon concentration and solid square is pH value.

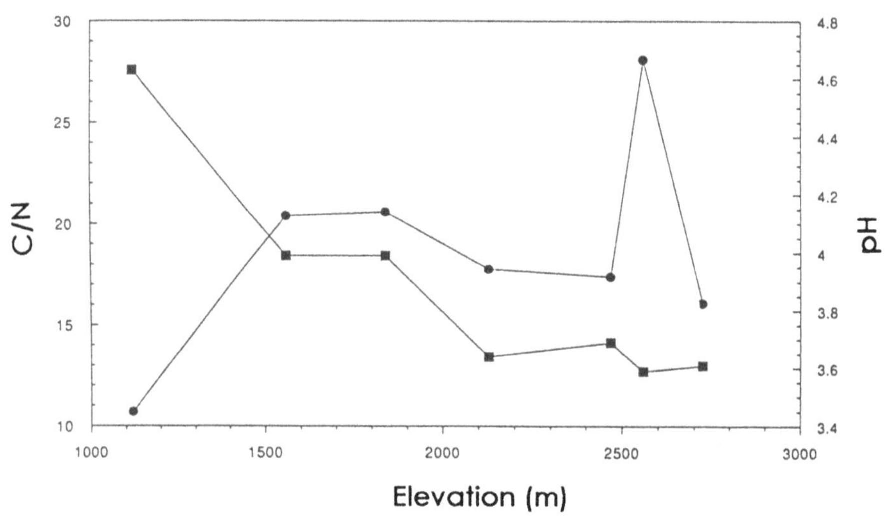

Fig. 6. pH value and C/N ratio of topsoils related to elevation of each experimental plot. Solid circle indicates C/N ratio and solid square is pH value.

elevation, that is the pH decreased as elevation increased. Therefore, Podzolic soils were distributed in the higher elevations, because of the strong acidic conditions of the topsoil there. Total carbon content of topsoil tended to increase with elevation,

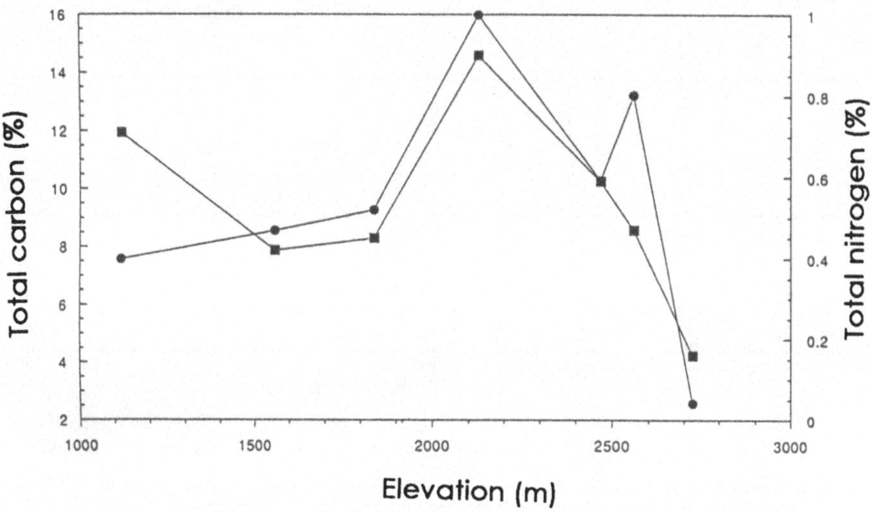

Fig. 7. Concentration of total carbon and total nitrogen of topsoils related to elevation of each experimental plot. Solid circle indicates total carbon and solid square is total nitrogen.

except in plot No. 7. This suggests that much organic matter is accumulated in the soil. These two facts, i.e., low pH values and considerable accumulation of organic matter in the topsoil, induce the formation of podzolic soils (Forest Soil Division, 1976). Therefore, changes in soil conditions of subalpine forests suggest that the increase in temperature accelerates the decomposition of soil organic matter and raises pH values (Poster and Post, 1986; Berg *et al.*, 1993).

3.3 Decomposition rate of *Tsuga diversifolia* litter

Temperature and water conditions affect litter decomposition, and high temperature and sufficient water accelerate litter decomposition (Jansson and Berg, 1985). The decomposition rate of needles and branches of *Tsuga diversifolia* was high at lower elevation (plot No. 1) and low at higher elevations (plot No. 6 and No. 7) (Fig. 8). Moreover, the rate of plot No. 3 and No. 4 was somewhat higher than that at plot No. 2. Air and soil temperatures had been already discussed in relation with elevation, indicating that the dependence of air and soil temperatures on the elevation is significantly affected by topography and vegetation. Therefore, the relationship between elevation and the decomposition rate did not indicate a clear negative correlation.

It remains to be proved that relationship between soil water condition and litter decomposition rate will be discussed more explicitly.

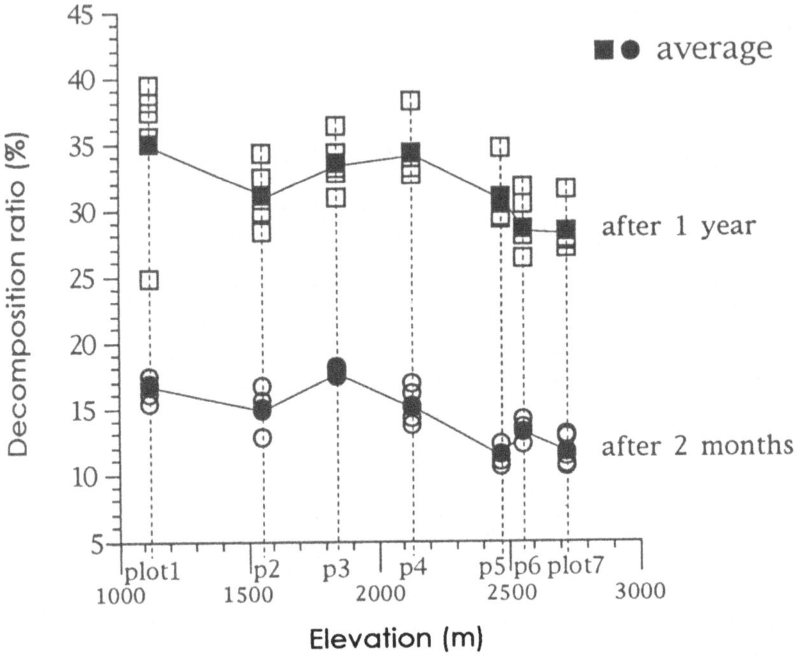

Fig. 8. Relationship between altitude and decomposition ratio of needles and branches of *Tsuga diversifolia*.

3.4 Climate change and site condition in the subalpine

The air and soil temperature conditions in alpine and subalpine zone have been discussed taking into account the freeze-thaw cycles and the formation process of microlandform (Sekine *et al.*, 1984; Higuchi, 1990; Matsuoka, 1991; Takahashi, 1995a,b). Although air temperature decreases nearly linearly at the lapse rate with the elevation, its decreasing tendence fluctuated depending on season in this study (Figs. 2, 3). Air and soil temperature conditions are affected by direct sun light(Higuchi, 1990), snow fall, vegetation, soil water condition, organic matter and landform (Sekine *et al.*, 1984; Takahashi, 1995a,b). The results of this study showed the fact that soil temperature of plot No. 6 where is windy and has lower and sparse forests as lower than that of plot No. 7 with higher elevation.

Air and soil temperature and soil fauna and flora affect litter decomposition rate (Takahashi, 1995a,b). Among of these factors, temperature is most important. As already described above, air and soil temperatures decreased with increasing the elevation. Therefore, litter decomposition rate decreased gradually with increasing the elevation. We found that pH in the topsoils (Fig. 5) decreases with the elevation, affecting the litter decomposition rate. Quite similar dependence of pH on the elevation was reported by Ohsumi and Kumada (1971) at Japanese North Alps and by Nakata *et al.* (1994) at Mt. Dairoku, Hokkaido. Values of pH also decreased with the elevation in the alpine of the Beartooth Plateau of Wyoming and Montana, U.S.A. (Johnson and Billings, 1962). It is considered that air and soil temperatures decrease with elevation and litter decomposition rate also decreases according increasing elevation (Schinner, 1982). According these facts, organic matter accumulated at higher elevation and soil water and soil became acidic.

If the global warming occurs, litter decomposition rate will be accelerated at higher elevation in subalpine. Therefore, the acidity of topsoil will become more neutral and then site condition of subalpine (Akaishi Mountains) will change, habitat of *Tuga diversifolia* which distributes at the ridge of podzolic soil will change and *Abies* and other species will invade this habitat.

4. CONCLUSIONS

(1) Air and soil temperatures decreased with increasing the elevation. However, this trend was different among each month and did not show linear changes because of effects of topography and vegetation.
(2) pH values in the topsoil decreased with increment in the elevation. Total carbon content in the topsoil tended to increase with the elevation. Soil conditions in higher subalpine forests were characterized by lower values of pH and the high accumulation of organic matter.
(3) The decomposition rate of twigs and leaves of *Tsuga diversifolia* was higher at low elevations and lower at high elevations. It remains to be proved that relationship between soil water condition and decomposition rate will be discussed more explicitly.

REFERENCES

Berg, B., McClaugherty, C. and Johansson, M-B., 1993. Litter mass-loss rates in late stages of decomposition at some climatically and nutritionally different pine sites. Long-term decomposition in a Scots pine forest VIII. *Can. J. Bot.*, **71**, 680-692.

Conacher, A. and Dalrymple, J.B., 1977. The nine unit landsurface model: An approach to pedogeomorphic research. *Geoderma*, **18**, 1-154.

Dalrymple, J.B., Blong, R.J. and Conacher, A.J., 1968. A hypothetical nine unit landsurface model. *Z. Geomorphol. N. F.*, **12**, 60-76.

EPA, IBAMA and USP (eds.), 1990. *Tropical Forestry Response Options to Global Climate Change*. IPCC, AFOS, 531pp.

Forest Soil Division, 1976. Classification of forest soil in Japan (1975). *Bull. Gov. Exp. Sta.*, **280**, 1-28. (in Japanese with English summary).

Higuchi, M., 1990. Fluctuation of air and soil temperatures through freezing and thawing periods on the bare ground of the top slope of Mt. Hoo. *Geogr. Rev.*

Jpn., **63**, 154-165 (in Japanese).

Jansson, P.-E. and Berg, B., 1985. Temporal variation of litter decomposition in relation to simulated soil climate. Long-term decomposition in a Scots pine forest V. *Can. J. Bot,.* **63**, 1008-1016.

Johnson, P.L. and Billings, W.D., 1962. The alpine vegetation of the Beartooth Plateau in relation to cryopedogenic processes and patterns. *Ecol. Monogr.*, **32**, 105-135.

Kobayashi, S., Morisada, K., Arimitsu, K. and Ohsumi, Y., 1984. Relationships between landforms and soil properties along to Norogawagoe-Toposequence. *Nichirinron*, **95**, 199-202. (in Japanese).

Kobayashi, S., Morisada, K., Ohsumi, Y. and Arimitsu, K., 1987. Effects of understory vegetation, soil, and microlandform on natural regeneration of subalpine forest in Akaishi Mountains, Japan. *Proceedings of Human Impacts and Management of Mountain Forests*, 373-385.

Maeda, T. and Miyakawa, K., 1970. *Evaluation of Plantation Site using Understory Vegetation Type.* Japanese Forestry Technical Association, Tokyo. 89pp. (in Japanese).

Matsuoka, N., 1991. A whole year ground temperature measurement in an alpine environment of Akaishi range, Japan. *Trans. Jpn. Geomorphol. Union*, 12-1, 41-49. (in Japanese with English summary).

Nakata, M., Tanaka, H. and Yagi, H., 1994. Altitudinal changes in vegetation and soils on Mt. Dairoku, Central Hokkaido, Japan. *Jpn. J. Ecol.*, **44**, 33-47 (in Japanese with English summary).

Ohsumi, Y. and Kumada, K., 1971. The study of alpine soil (I). *Soil Sci. Plant Nutr.*, **42**, 45-51 (in Japanese).

Poster, J. and Post, W.M., 1986. Influence of climate, soil moisture, and succession on forest carbon and nitrogen cycles. *Biogeochemistry*, **2**, 3-27.

Schinner, F., 1982. Soil microbial activities and litter decomposition related to altitude. *Plant Soil*, **65**, 87-94.

Schlesinger, W.H., Reynolds, J.F., Cunningham, G.L., Huenneke, L.F., Jarrell, W.M., Virginia, R.A. and Whitford, W.G., 1990. Biological feedbacks in global desertification. *Science*, **247**, 1043-1048.

Sekine, K., Takeuchi, K., Kato, E. and Tazoe, Y., 1984. Annual and diurnal changes of soil temperature on the Mt. Tateyama and the Ontake volcanoes. *Geogr. Rev. Jpn.*, **57**, 663-675. (in Japanese with English summary).

Takahashi, M., 1995a. *The Study of Nutrients Regime in Organic Matter and Its Management.* Doctor thesis of Hokkaido University, 157pp.

Takahashi, N., 1995b. Distribution of the autumn ground temperature and its controlling factors in the alpine zone of the central Daisetsuzan mountains. *Geogr. Rev. Jpn.*, **68**, 27-42. (in Japanese with English summary).

MODELING OF SNOWMELT AND SOIL TEMPERATURE EVOLUTION IN SUBALPINE SNOW PATCH

Yoshikazu Ohtani[1], Takeshi Morisawa[2], Katsumi Yamanoi[1], Hiromu Daimaru[3] and Yoshiaki Goto[4]

ABSTRACT: Micrometeorological models which describe snowmelt process and the soil temperature evolution after snowmelt were reported in this study. The snow patch which we studied was located on Mt. Zarumori, Akita Prefecture, Japan. Parameters used for the model were estimated and confirmed by heat balance observation. In addition, snowfall distribution and general meteorological observations were conducted around the site, and used for the model input. The model was verified by comparing the calculated values with the observed values. The model was able to predict the shift of the snow edge within a temporal error of about 3 days and a spatial error of 5m. The calculated and observed soil temperatures agreed well, and it appeared that the calculated value of accumulated temperatures related to the period of time after the snow released. The phenological stages of the snow patch vegetation changed with the increase of the estimated effective accumulated temperature.
KEY WORDS: modeling, snow patch, effective accumulated temperature, phenology.

1. INTRODUCTION

Japanese snow patch vegetation is one of the typical vegetation types found in alpine and subalpine areas covered with deep snow. In snow patches formed by snowdrifts, the spatial distribution of the snow depth is mainly affected by the interaction between topography and the prevailing wind direction in winter. A ring-shaped distribution of vegetation according to the snow release has been reported (Hayashi, 1935). These phenomena have relationships between temporal changes in micro-environments and the phenology of the vegetation (Billings and Bliss, 1959; Mizuno, 1984).

---

[1]Forestry and Forest Products Research Institute, P.O. Box 16, Tsukuba Norin, Ibaraki 305, Japan.

[2] Kiso Experimental Station, Forestry and Forest Products Research Institute, Kisofukushima, Nagano 397, Japan.

[3] Tohoku Research Center, Forestry and Forest Products Research Institute, Morioka, Iwate 020-01, Japan.

[4] Kansai Research Center, Forestry and Forest Products Research Institute, Kyoto 612, Japan.

From both ecological and micrometeorological viewpoints, it is important to predict the future changes in alpine vegetation induced by global warming. The approach that clarifies a quantitative connection between the phenology of vegetation and the micrometeorological environment is most appropriate for this purpose.

In this paper, a micrometeorological model was developed for the snowmelt process and soil temperature evolution after snow release. Differences between the calculated and observed data are discussed. Experimental study on the prediction of phenological stages of snow patch vegetation also was conducted using the concept of effective accumulated temperature.

2. STUDY SITE

The snow patch was on Mt. Zarumori, Akita Prefecture (39°46'N, 140°48'E), at an elevation of 1540m (Fig.1). On the lee side of this mountain, snow drift forms large snow patches every year. Daimaru and Ikeda (1993) have reported that the snow patches sometimes remain until late July.

There are three types of patchy vegetation on the lee slope: a *Sasa kurilensis* community, an *Alnus crispa* community, and a snow patch community. The snow patch community consists mainly of herbs such as *Fauria crista-galli*, *Calamagrostis purpurascens* subsp. *arctica*, *Primura nipponica*, *Sanguisorba albiflora* and *Hosta rectifolia*. It also contains a few woody plants such as *Phyllodoce aleutica*. The vegetation on the slope where the shift of the snow patch was observed was *Primura nipponica-Fauria crista-galli* Association (Morisawa *et al.*, 1993). This is typical snow patch vegetation of northeastern Japan (Ishizuka, 1949).

3. OBSERVATION

For both snowmelt and soil temperature models, meteorological observations were

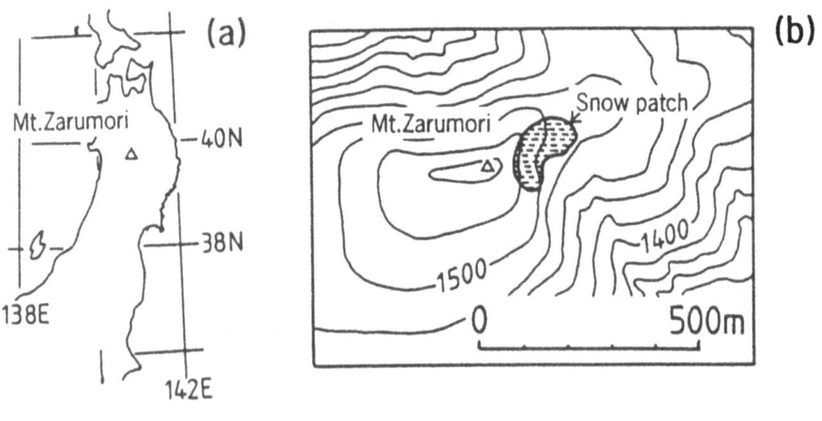

Fig. 1. Location of Mt. Zarumori (a), and the study site (b).

conducted at the top of Mt. Zarumori about 100m from the snow patch. Global solar radiation, air temperature, air humidity and wind speed were measured using a phyranometer (at 2.0m), a temperature and humidity probe with radiation shield (at 1.5m), and a cup anemometer (at 2.0m), respectively. Soil temperatures were measured at two different depths (5cm, 10cm) near the center of the snow patch to verify the soil temperature model.

A positional change of the upper edge of the snow patch was observed during the period of June 13 to July 20, at approximately 7 day intervals. These data were used to verify the model of snowmelt. The phenology of the typical species in the snow patch also was observed in the same period. These observations were continued after snowmelt until late August. Energy budget observations also were conducted in and around the same snow patch to determine bulk transfer coefficients correspond to the snow and the thin vegetation surfaces.

4. MODELING

4.1 Snow melt

Kondo and Yamazaki (1990) have proposed a snowmelt model by a heat balance method. The snowmelt model in this study is based on the same principle and used to estimate the date that snow disappears completely.

The heat budget above the snow surface is represented as follows:

$$(1-\alpha)S^{\downarrow} + L^{\downarrow} = L^{\uparrow} + H + lE + G_0 \quad (1)$$

$$Rn = (1-\alpha)S^{\downarrow} + \left(L^{\downarrow} + L^{\uparrow}\right) \quad (2)$$

$$G_0 = Rn - H - lE \quad (3)$$

where $S^{\downarrow}$ is the global solar radiation, α is the surface albedo, $L^{\downarrow}$ is the downward long wave radiation, $L^{\uparrow}$ is the upward long wave radiation, H is the sensible heat flux, lE is the latent heat flux, and G_0 is the heat flux into the snow. The energy of G_0 represents the phase or the temperature changes of the snow.

Expressing H and lE in the bulk formulations,

$$H=C_p \rho C_{hs} U(T_s-T) \quad (4)$$

$$lE=l \rho C_{es} U(q.(T_s)-q) \quad (5)$$

where Cp is the specific heat of air at a constant pressure and ρ is the air density. C_{hs} and C_{es} are bulk transfer coefficients for sensible and latent heat, respectively, above the surface. U is the wind velocity, T_s is the surface temperature, T is the air temperature, $q.(T_s)$ is the saturated specific humidity at the surface temperature T_s, and q is the specific humidity of the air.

The model consists of the snow-atmosphere energy transform process expressed by Eqs. (1) to (5), and the freezing, melting and temperature-changing processes of the snow. Global solar radiation, air temperature, specific humidity of the air, wind velocity and surface albedo are required as the input data. The downward long wave radiation is estimated by the Yamamoto-Brunt formula. Calculation of G_0 was achieved by applying measured data for the surface albedo and bulk transfer coefficients.

The heat conduction flux in the frozen snow layer(q_s) can be calculated from,

$$q_s = -\lambda_{dry}\left(\partial T_s / \partial z\right) \tag{6}$$

where λ_{dry} is the thermal conductivity of the dry snow, T_s is the snow temperature, and z is the depth of the snow. G_0 is q_s at the snow surface ($z = 0$) calculated from Eq. (6).

4.2 Soil temperature evolution

The soil temperature model was used to estimate the soil temperature rise after the snowmelt. On the thin vegetation surface, it is possible to deal with the energy exchange from the Eqs. (1) through (5), as well as the snow surface.

Heat conduction flux (q_g) in the soil layer is simply shown as,

$$q_g = -\lambda_g\left(\partial T_g / \partial z\right) \tag{7}$$

where λ_g is the thermal conductivity of the soil, T_g is the soil surface temperature, and z is the depth in the soil. G_0 is q_g at the soil surface calculated from Eq. (7). By coupling the heat transfer in the atmosphere and the soil, and joining Eqs. (1) through (5) and (7), both the heat budget of the soil and the soil temperature at different depths can be estimated.

4.3 Parameters used to run the model

The snowmelt process was simulated for the spring of 1993. The initial distribution of the snow depth was estimated based on the profile investigated in 1992, in which the upper edge of the snow patch was adapted to that of June 13, 1993. The meteorological data of global solar radiation, air temperature, air humidity, and wind speed were input to the model at intervals of 10 minutes. During the period of calculation, no new snowfall appeared at the site.

The following values of parameters were decided by the energy balance observation on the snow surface. The surface albedo was determined to be α=0.5, and the bulk transfer coefficients for the sensible and latent heat were confirmed to be C_{hs} = 0.002 and C_{es} = 0.002, from the result of Kondo and Yamazawa (1986). The dry snow density must be specified in the transformation from the snow melting energy to the depth. The estimation was to be 0.46 Mgm^{-3}, based on the measurement of wet snow density, that assumed a volumetric water content of 15%.

When the snow cover disappeared, the calculation of the soil temperature rise was initiated. The parameters used to run the soil temperature model were as follows. The surface albedo and the bulk transfer coefficients for the sensible and latent heat were decided to be α =0.2, C_{hg} = 0.03, and C_{eg} = 0.01, respectively, by the energy balance observation on the thin vegetation surface. Heat conductivity, specific heat and the density of the soil were estimated to be 0.834 $Wm^{-2}K^{-1}$, 2.81 $kJkg^{-1}K^{-1}$ and 0.85 Mgm^{-3}, respectively, from the soil temperature measurements assuming the three-layer distribution of the typical subalpine marsh.

5. RESULTS AND DISCUSSION

5.1 Snow melt process

The temporal changes in the observed upper edge of the snow patch and the calculated snow profile are shown in Fig.2. The observed upper snow edge receded more rapidly than calculated by the model. The snow patch completely disappeared on July 21, at +4m (indicated by the filled triangle in Fig.2). Disappearance of the snow patch edge was estimated by the model to occur on July 18, at 0m. The temporal differences was three days, while the spatial difference was 4m. There was a possibility that the error in the estimation of the initial snow profile influenced the result. However, Higuchi *et al.* (1979) reported that the snow patches formed by the snow drifts only had minor changes inter-annually in their shapes. Considering the temporal and the spatial differences of the phenology of the snow patch vegetation, the error seemed to be acceptable.

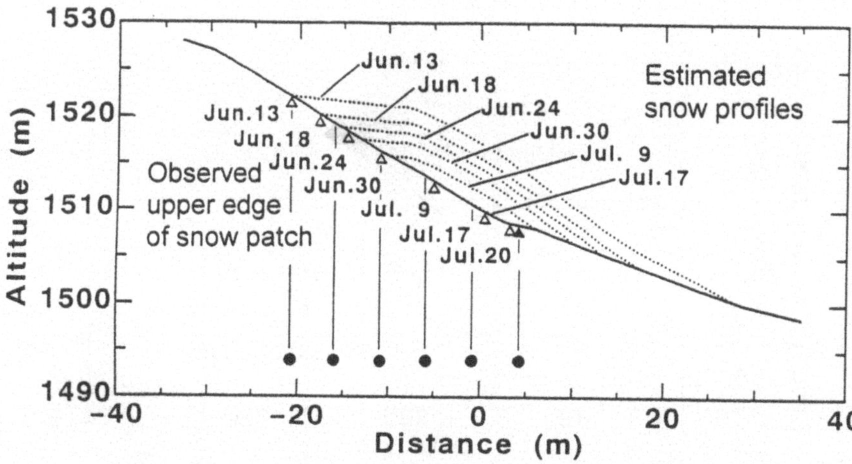

Fig. 2. Temporal changes of the upper boundary of snow. Observed locations of the upper edge of the snow patch are shown with triangles, and the calculated snow profiles are indicated with dotted lines. The filled triangle shows the point where the snow patch disappeared, and the filled circles show the location on the slope where effective accumulated temperature is estimated.

5.2 Soil temperature evolution

The process of the soil temperature rise after the snowmelt was explained by the effective accumulated temperature. The concept of the effective accumulated temperature is often used in relation to the plant production. The estimated and the observed mean daily soil temperatures at the 5cm depth agreed well. The effective accumulated soil temperature (T_a) was calculated by the following equation,

$$T_a = \sum\left(T_m - T_b\right) \tag{8}$$

where T_m is the estimated daily mean soil temperature, and T_b is the standard temperature set to be 0℃ after Callaghan (1974).

Expected effective accumulated soil temperatures in the different locations are shown in Fig.3. At the beginning of the curve, the small gradient parts are seen at -21m and -16m. This tendency may be related to the time after the snow patch disappeared. Therefore, snow cover influences soil temperature evolution. The snow cover holds the soil surface temperature at 0℃ until the snow disappears. If the global warming only influences temperature, the temperature rise may increase the effective accumulated temperature as follows: 1) snowmelt is accelerated and the starting time of the effective temperature accumulation shifts forward; 2) daily mean temperature rises, increasing the inclination of the accumulated temperature curve.

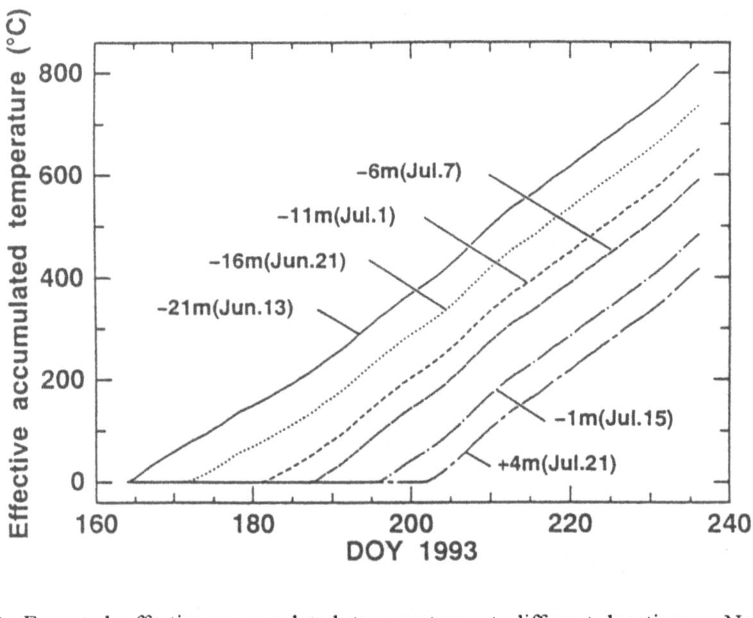

Fig. 3. Expected effective accumulated temperature at different locations. Numerical values represent the position on the slope (shown in Fig.2). The date snow disappeared is shown in parentheses. Day of year (DOY) represents number of days counted from January 1.

The temperature resource for the snow patch vegetation may be increased rapidly by the reasons listed above.

5.3 Effective accumulated temperature and the phenology of *Faulia crista-galli*
The relationship between the calculated effective accumulated temperature and the phenological stages of *Faulia crista-galli*, that is one of the dominant species in the snow patch, is shown in Fig.4. It appears that the *Faulia crista-galli* changes its phenological stages with an increase in the effective accumulated temperature. Furthermore, there is a tendency for the phenological stages to progress faster in closer proximity to the snow patch.

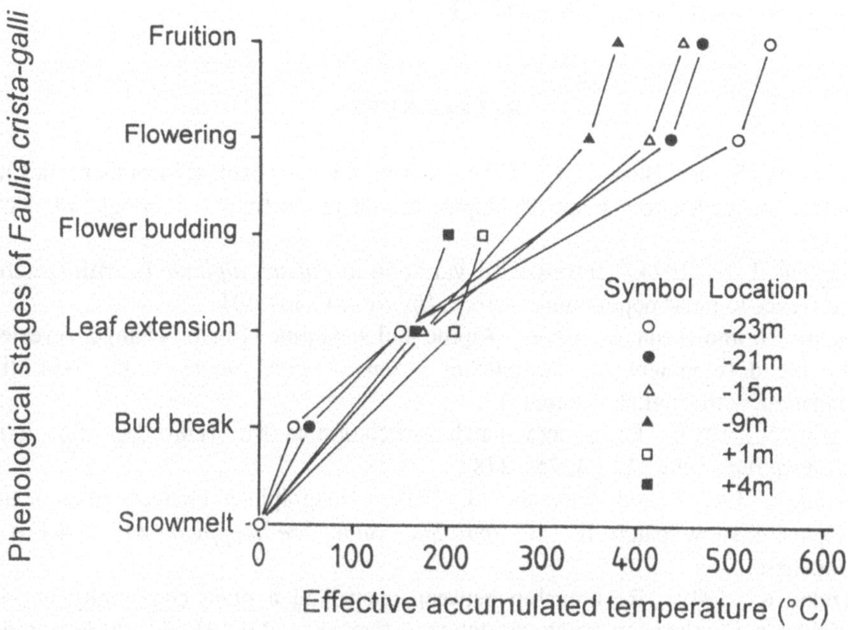

Fig. 4. Relationship between the effective accumulated temperature and the phenological stages of *Faulia crista-galli* at different locations on the slope.

6. CONCLUSIONS

Micrometeorological models which describe snowmelt and the soil temperature evolution after snowmelt were made in this study. The model was verified by comparing the calculated values with observed values. The model was able to predict the movement of the boundary of the snow according to the release from the snow cover within a temporal error of three days and a spatial error of 5m. The calculated and observed soil temperatures agreed well, and it seemed that the trends of calculated accumulated temperatures related to the time after the snow disappeared. The phenological stages of *Faulia crista-galli* changed with the increase of the effective accumulated temperature. There was a tendency for the phenological stages to progress faster in closer proximity to the snow patch.

ACKNOWLEDGMENT

We wish to thank Shigeto Ikeda and Hiromichi Onodera of the Tohoku Res. Center of FFPRI, Yasuo Osumi of JIRCAS, and Yasuko Mizoguchi of FFPRI for their help in setting up the experimental plot and in collecting data on site. We would also like to thank Shuuichiro Yoshinaga of FFPRI for providing the physical properties of the alpine marsh soil. Finally we are pleased to acknowledge the helpful advice of Hiroshi Taoda and Shigeo Kobayashi of FFPRI.

REFERENCES

Billings, W. D. and Bliss, L. C., 1959. An alpine snowbank environment and its effects on vegetation, plant development, and productivity. *Ecology*, **40**, 388-397.

Callaghan, T. V., 1974. Intraspecific variation in *Phleum alpinum* L. with specific reference to polar populations. *Arct. Alp. Res.*, **6**, 361-401.

Daimaru, H. and Ikeda, S., 1993. Alpine and subalpine climatic change, revealed by the development of snowpatches. *Jpn. J. For. Environ.*, **35**, 9-14. (in Japanese with English summary)

Hayashi, N., 1935. On a snow-patch association at Mt. Hakkoda. *Sci. Rep. Tohoku Imp. Univ., Ser.* 4, 253-278

Higuchi, K., Ota, T. and Watanabe, O., 1979. Geographical characteristics of the Japanese snow patch II. *J. Jpn. Soc. Snow Ice (Seppyo),* **41**, 77-84. (in Japanese)

Ishizuka, K., 1949. Relationship between snow and a plant community in Mt. Hakkoda. On the snow patch vegetation. *Ecol. Res.*, **11**, 193-203. (in Japanese)

Kondo, J. and Yamazaki, T., 1990. A prediction model for snowmelt, snow surface temperature and freezing depth using a heat balance method. *J. Appl. Meteorol.*, **29**, 375-384.

Kondo, J. and Yamazawa, H., 1986. Bulk transfer coefficient over a snow surface. *Bound.-Layer Meteorol.*, **34**, 123-135.

Mizuno, K., 1984. Habitat types of Ohanabatake vegetation in the Akashi mountains, central Japan. *Geogr. Rev. Jpn.*, **57**(Ser.A)-6: 384-402. (in Japanese with English summary).

Morisawa, T., Goto, Y., Ohtani, Y. and Daimaru, H., 1993. Phenology in a snowpatch (I) A case study at Mt. Zarumori, Akita Prefecture. *Trans. Jpn. For. Soc.*, **104**, 755-756. (in Japanese)

EXPANSION OF ELEVATIONAL DISTRIBUTION OF BEECH (*Fagus crenata* Blume) ALONG THE CLIMATIC GRADIENT FROM THE PACIFIC OCEAN TO THE SEA OF JAPAN IN HONSHU, JAPAN

Nobuyuki Tanaka[1] and Hiroshi Taoda[1]

ABSTRACT: There is a climatic gradient of snowfall between the Pacific Ocean side and the Sea of Japan side of the Honshu island as well as a temperature gradient along latitude. In summer, precipitation is plentiful on the both sides. In winter, however, there is a little precipitation in the Pacific Ocean side whereas there is a lot of snowfall in the Sea of Japan side. The elevational range of beech distribution was determined in areas around 37°N and 35°20'N on the Honshu island. The upper and lower limits of beech distribution are 300 m higher and 250 m lower in the Sea of Japan side than in the Pacific Ocean side, respectively. These results suggest that the beech distribution on a macro-scale is affected by snowcover as well as by temperature.
KEY WORDS: forest zones, temperature, seasonal pattern of precipitation, snowcover, climatic change.

1. INTRODUCTION

Temperature is the climatic factor primarily determining the forest zonation (latitudinal and elevational) in Japan, where the precipitation is sufficient for plant growth. However, there is a difference in seasonal patterns of precipitation between the Pacific Ocean side and the Sea of Japan side. In summer, precipitation is plentiful on the both sides. In winter, however, there is a little precipitation in the Pacific Ocean side whereas there is a lot of snowfall along the Sea of Japan side. This means that there is a climatic gradient of precipitation seasonality between the Pacific Ocean side and the Sea of Japan side, as well as a temperature gradient along latitude.

There is also a difference in the floristic composition of natural forests in the Pacific Ocean side and the Sea of Japan side. In the montane (cool-temperate) zone, for example, beech (*Fagus crenata* Blume) is a leading dominant in the Sea of Japan side, whereas various deciduous broad-leaved and evergreen coniferous tree species, mixed with beech, dominate in the Pacific Ocean side.

Recent climatological studies have warned of rapid global warming due to

[1] Forestry and Forest Products Research Institute, P.O. Box 16, Tsukuba Norin, Ibaraki 305, Japan.

massive emissions of greenhouse gases such as CO_2 and CH_4 from modern industries. In order to predict the effects of climate change on plant distribution, it is necessary to clarify the main factors governing the present distribution of each plant species.

A comparison of the elevational zonation of vegetation and climate in the mountains along the Pacific Ocean-Sea of Japan gradient is important to clarify the causes and processes of this zonation. This study investigated the upper and lower limits of beech and climatic factors to characterize the elevational distribution of beech along this gradient.

This paper is a revision of a preceding paper (Tanaka, 1981). Nomenclature follows Kitamura and Murata (1971, 1979).

2. STUDY AREA

Regions along 37°N latitude were selected for the study since there are contiguous mountains from Abukuma and Yamizo near the Pacific Ocean coast through Oku-nikko and Oze in the midlands to Echigo near the Sea of Japan coast. A region near the Pacific Ocean coast around 35°20'N was also selected since there is no high mountain reaching the subalpine (subarctic) zone in the Abukuma and Yamizo Mountains (Fig. 1).

As mentioned earlier, there is a climatic gradient of snowfall between the Pacific Ocean side and the Sea of Japan side. The annual maximum snow depth (MSD) is less than 50 cm in the Pacific side. but it increases toward the mountainous areas in the Sea of Japan side, where it reaches more than 400 cm (Fig. 1). Walter s climatic diagrams of three sea-side cities, Takada (37°06'N) near the Sea of Japan, and Onahama (36°57'N) and Mishima (35°07'N) near the Pacific Ocean show a clear difference between the two sides (Figs. 1 and 2). These climatic conditions are determined by the seasonal change of dominant air masses around the Japanese Arichipelago. In winter, a cold Siberian air mass blows over the warm Sea of Japan Current and brings about heavy snowfall on the Sea of Japan side, but after going over the central mountain range it becomes dry and blows over the Pacific side. In summer, warm North Pacific air masses and typhoons bring about rainfall on the both sides of the Honshu island.

3. METHODS

The author determined the upper limit elevation of beech as well as other dominant tree species by recording component tree species of forests along trails in the mountains at 37°N, that is, Nantai-san (summit elevation: 2,484 m a.s.l.), Aizu-komaga-take (2,132 m), Arasawadake (1,969 m), and Nakanodake (2,085 m) and in those at 35°20'N, that is, Tekaridake (2,591 m) and Fuji-san (3,776 m).

In order to determine the lower limit of beech distribution, it is necessary to find natural and aged forests since beech as a climax species is replaced by other seral species in secondary forests which have been frequently felled in the past. Well-

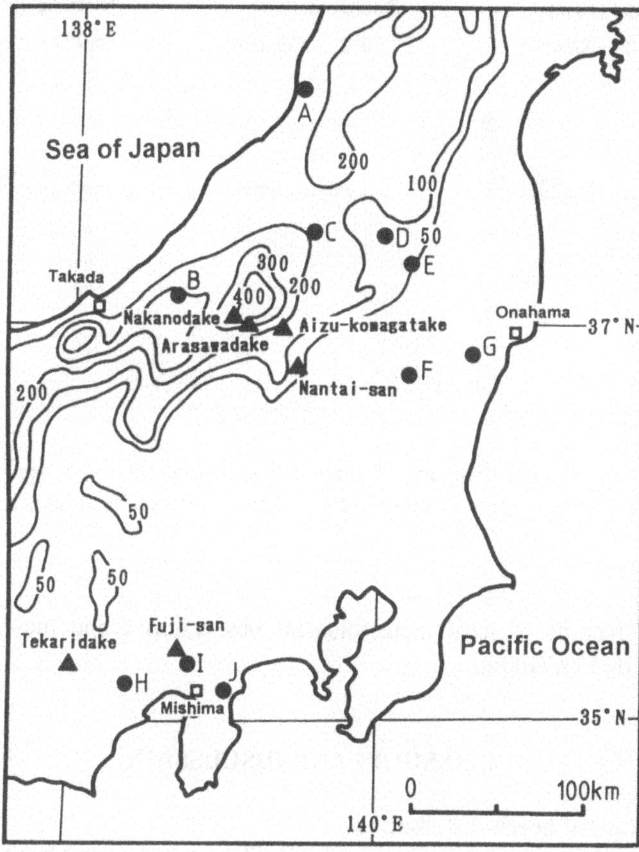

Fig. 1. Study area and annual maximum snow depth (cm). Solid triangles and circles denoted by letters (A to J) indicate mountains, and forest stands that contain beech, respectively.

preserved forests are fragmented and limited in the lower part of beech distribution, where human activities have had more influence on the vegetation. The author identified forests containing beech at the lowest elevation in each studied region from the Environment Agency's "Important Plant Communities in Japan (1980)", which describes many preserved vegetation.

Kira's warmth index (WI) well delineates the boundaries of the forest zones. For example, the values of 85 and 45 (°C·month) WI indicate the boundaries between zones of warm-temperate evergreen broad-leaved, cool-temperate deciduous broad-leaved and subalpine evergreen coniferous forest, respectively (Kira, 1948). The WI value for each site is calculated from WI $=\Sigma(t_1 -5)$, where t_1 is the mean monthly temperature exceeding 5°C. I estimated the boundary elevation, using

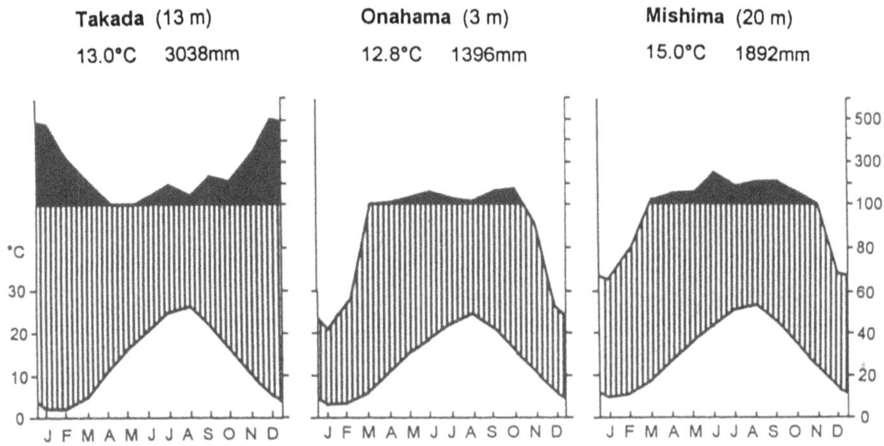

Fig. 2. Walter's climatic diagrams of three cities, Takada near the Sea of Japan, and Onahama and Mishima near the Pacific Ocean. The location of the cities is shown in Fig.1.

temperature records of local meteorological observatories and the lapse rate of temperature (0.55°C/100m).

4. RESULTS AND DISCUSSION

4.1 Upper limit of beech distribution

The annual maximum snow depth (MSD), which corresponds to winter precipitation, can indicate the rating of the Pacific-Sea of Japan gradient (Fig. 1). Around 35°20'N, Tekaridake and Fuji-san, which are in the area with MSD less than 50 cm, are in the Pacific climate. Along 37°N, MSD increases successively from Nantai-san (about 50 cm MSD), through Aizu-komagatake and Arasawadake to Nakanodake (over 400 cm MSD).

The elevational distribution of dominant tree species in cool-temperate and subalpine zones of each surveyed mountain is shown in Fig. 3, where the boundary elevations of the two zones at 37°N and 35°20'N are adjusted, based on the elevation of 45 WI, that is the value between subalpine and cool-temperate zones.

On Tekaridake, the upper limit of beech is 1,720 m a.s.l. while the lower limit of subalpine fir species (*Abies mariesii* Mast. and *A. veitchii* Lindl.) is about 1,750 m a.s.l. Along the Shojin trail on the northern slope of Fuji-san, the upper limit of beech and cool-temperate fir (*A. homolepis* Sieb. et Zucc.) almost coincides with the lower limit of *A. veitchii* at 1600 m a.s.l. The boundary elevation between cool-temperate and subalpine zones estimated by 45 WI is about 1,700 m around 35°20'N. This elevation coincides with the upper limit of beech in Tekaridake and is 100 m higher than that at Fuji-san. Another feature of these mountains is the presence of

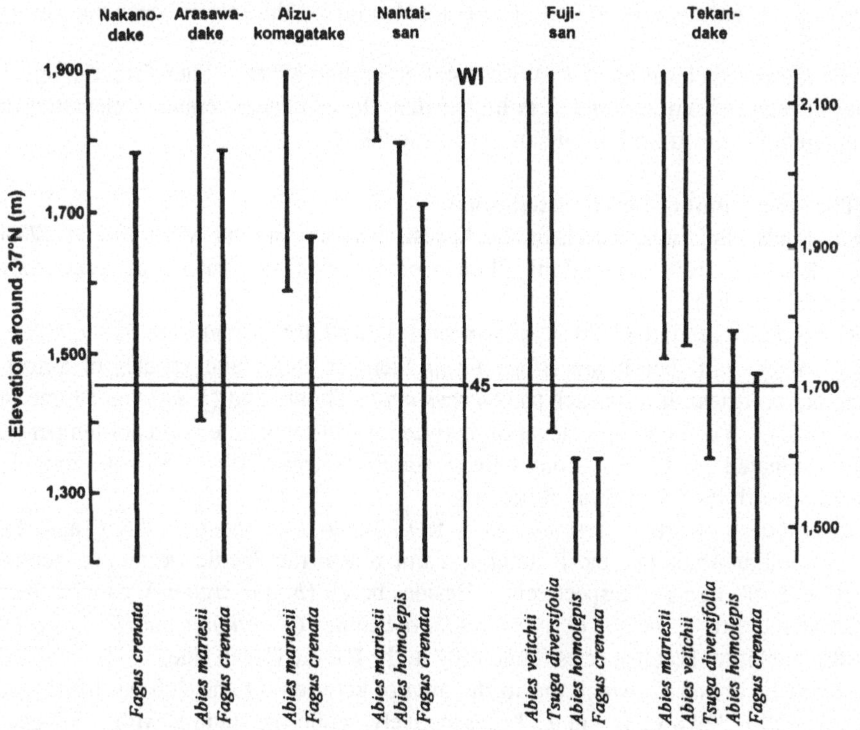

Fig. 3. Elevational distribution of main tree species between the upper cool–temperate and the lower subalpine zones.

a hemlock zone, where *Tsuga diversifolia* Masters dominates in the lower part of subalpine zone.

On Nantai-san, the upper limits of the beech and *A. homolepis* are 1,710 and 1,800 m a.s.l., respectively, while the lower limit of subalpine fir (*A. veitchii* and *A. mariesii*) connects to the upper limit of *A. homolepis*. On Aizu-komagatake, the upper limit of beech is 1,670 m and the lower limit of *A. mariesii* is 1,590 m a.s.l. There are mixed beech-fir forests in an 80 m elevational range of overlapping distribution. The hemlock zone which is prominent on mountains in the Pacific side is not clear on Nantai-san and is not found on Aizu-komagatake.

On Arasawadake, there is a 360 m elevational range of overlapping distribution with a beech upper limit of 1,770 m a.s.l. and an *A. mariesii* lower limit of 1,410 m a.s.l. However, unlike Aizu-komagatake, there are no mixed beech-fir forests. The distribution of beech and fir is segregated topographically as fir trees grow only on flat plateaus and beech trees grow on slopes and ridges. In the upper limit area of beech, there are beech thickets of Krummholz, sometimes mixed with dwarf oak (*Quercus mongolica* Fischer ex Turcz. var. *undulatifolia* (Leveil.) Kitamura et

Horikawa), a main component species of the pseudo-subalpine zone in Japan.

On Nakanodake, the upper limit of beech is 1,770 m a.s.l., being the same as Arasawadake, but there is no fir. There are also mixed thickets of beech and dwarf oak.

The estimated elevation of 45 WI is 1,450 m around 37°N. Therefore, the upper limit of beech is from 220 to 320 m higher than the estimated boundary elevation in the mountain series from Nantai-san to Nakanodake.

4.2 The lower limit of beech distribution

Forest stands containing beech at the lowest elevation around 35°20'N and 37°N were identified from "Important Plant Communities in Japan", as previously mentioned (Table 1).

Three stands around 35°20' N, shown as H, I and J are located from 620 to 880 m a.s.l. Mixed with beech are *Abies firma* Sieb. et Zucc. and species of warm-temperate evergreen forests such as *Quercus acuta* Thunb. and *Q. salicina* Blume in these stands. The boundary elevation between warm-temperate and cool-temperate zones estimated by 85 WI around these stands is from 700 to 800 m, roughly coinciding with the lower limit of beech.

Seven beech stands represented as A to G are located around 37°N (Table 1). The G stand at Abukuma and F stand at Yamizo near the Pacific Ocean are located at 650 and 400 m a.s.l., respectively. Besides beech (*Fagus crenata*), canopy trees of the stands are *A. firma*, *Q. myrsinaefolia* Blume, *Q. salicina* and *F. japonica* Maxim., the other beech species found only in the Pacific Ocean side.

Stands E, D and C, which are in the area of between 50 and 200 cm MSD, are located at 540, 520 and 440 m a.s.l., respectively, while the stand B with MSD over 300 cm is located at 220 m a.s.l. The stand A in Murakami-shi, Niigata Prefecture is located on a lowland with 40 m a.s.l. The estimated elevation of 85 WI around 37°N is from 400 to 500 m a.s.l. Therefore the lower limit of beech is from 200 to 300 m lower than the boundary elevation estimated by 85 WI.

4.3 Elevational distribution range of beech

The detail analysis of temperature conditions and distribution of Japanese conifers revealed that most of the species showed normal curves of distribution density along the WI gradient (Kira and Yoshino, 1967). Such analysis is not available for broad-leaved tree species due to the lack of detail data on their overall distribution. Kira (1976) examined temperature conditions of the 12 upper and 11 lower limit sites of beech distribution and concluded that they roughly corresponded to the thermal conditions of 45 and 85 WI, respectively. However, the changes in beech distribution along the climatic gradient between the Pacific Ocean side and the Sea of Japan side can not be explained by only WI index.

Based on the data on the upper and lower limits of beech distribution discussed in this paper, the elevational distribution of beech along the Pacific Ocean-Sea of Japan gradient is shown in Fig. 4, where the elevational descent of forest zones from 35°20'N to 37°N due to the thermal decrease is adjusted, using the WI index in order to compare the forest zonation of different latitudes. The thermal conditions at 1700 m (45 WI) and 600-800 m a.s.l. (85 WI) around 35°20'N are equivalent to

Table 1 Forest stands containing beech in the lowest elevations

Stand	Elevation (m)	Main overstory species	Main understory species
A	40	*Fagus crenata* *Tilia maximowicziana* *Carpinus cordata* *Quercus mongolica* var. *grosseserrata* *Carpinus laxiflora*	*Sasa palmata** *Aucuba japonica* var. *borealis**
B	220	*Fagus crenata* *Magnolia obovata*	*Magnolia salicifolia* *Viburnum furcatum* *Sasa palmata**
C	440	*Cryptomeria japonica** *Fagus crenata*	*Aucuba japonica* var. *borealis**
D	520	*Fagus crenata* *Pinus densiflora** *Quercus mongolica* var. *grosseserrata*	*Skimmia japonica* var. *intermedia** *Tripetaleia paniculata*
E	540	*Fagus crenata* *Carpinus laxiflora*	*Rhododendron wadanum* *Rhododendron semibarbatum*
F	400	*Fagus japonica* *Fagus crenata* *Acer mono* *Quercus myrsinaefolia** *Quercus salicina** *Abies firma**	
G	650	*Fagus crenata* *Quercus mongolica* var. *grosseserrata* *Abies firma** *Fagus japonica*	*Sasamorpha borealis**
H	620	*Fagus crenata* *Stewartia monadelpha* *Abies firma** *Quercus salicina**	*Sasamorpha borealis**
I	880	*Fagus crenata* *Quercus mongolica* var. *grosseserrata* *Acer mono*	*Sasamorpha borealis**
J	680	*Quercus acuta** *Fagus crenata* *Stewartia monadelpha* *Zelkova serrata*	*Aucuba japonica** *Pleioblastus chino* var. *gracilis**

Stand location, see Fig.1. * : Evergreen species.

those of 1400-1500 m and 400-500 m a.s.l. around 37°N, respectively. This diagram illustrates that the upper and lower limits of beech distribution are 300 m higher and 250 m lower in the Sea of Japan side than in the Pacific Ocean side, assuming that the upper and lower boundaries of the cool-temperate zone are at the elevation of 45 and 85 WI, respectively. Thus, the elevational range of beech distribution expanded upward and downward along the climatic gradient from the Pacific Ocean side to the Sea of Japan side. The elevational range of beech distribution is from 900 to 1,100 m in the Pacific Ocean side, whereas it is from 1,500 to 1,600 m in the Sea of Japan side. This distribution pattern indicates that

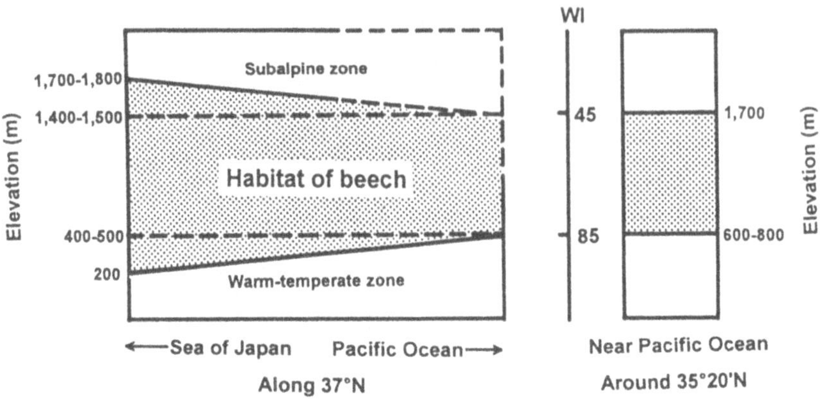

Fig. 4. Elevational distribution of beech along a cross-section from the Pacific Ocean to the Sea of Japan. The shaded areas indicate the habitat of beech. The three elevational zones delineated by the elevations of 45 and 85 of WI are the subalpine, cool-temperate and warm-temperate zones, in descending order. WI: Kira's warmth index.

the range of beech habitat is affected by snowcover as well as by temperature. Although winter is the dormant season for beech as a summer-green tree species, snowcover may help beech seeds, seedlings and trees survive and grow in the moist ground in spring and early summer.

In addition to the distribution, snowcover may also favor the dominance of beech in the Sea of Japan side. Beech dominates in natural forests in the Sea of Japan side, whereas it shares the dominance with many other tree species, such as *Quercus mongolica* var. *grosseserrata*, *Acer mono* Maxim., *Carpinus* spp. and *Abies firma*, in the Pacific zone. It seems that beech reproduction is more favored by snowcover than other tree species which exist more densely in the Pacific Ocean side with dry winters.

4.4 Influence of climatic change on beech distribution

This study suggests that beech distribution on a macro-scale is mainly affected by snowcover and temperature. In order to predict effects of climatic change on beech distribution, it is necessary to obtain information on seasonal patterns of precipitation and temperature, which are used to estimate the values of MSD and WI.

In view on the time scale of more than 100 years, the present pattern of beech distribution may shift upward or downward in response to the temperature change, if the precipitation conditions do not change. On the other hand, changes in winter precipitation may affect beech distribution, even if the temperature conditions do not change. In the Pacific Ocean side, an increase in snowfall may assist beech to expand its habitat along elevation and increase its dominance in forests. In the Sea of Japan side, a decrease in snowfall may reduce its elevational distribution and dominance in forests.

5. CONCLUSION

The upper and lower limits of beech distribution coincide with the boundary elevations determined by 45 and 85 WI in the Pacific Ocean side, but extend 300 m higher and 250 m lower than those elevations in the Sea of Japan side, respectively. This study suggests that beech distribution on a macro-scale is affected by both snowcover and temperature. Snowcover apparently assists the expansion of elevational distribution of beech in the Sea of Japan side.

In addition to the climatic factors which control the vegetation on a macro-scale, edaphic factors such as those of geology, topography and soil are also important on a meso-scale. A study on beech habitat in relation to the elevation, topography and soil of Mt. Hiuchidake showed that the upper limit of beech distribution is higher on steep slopes than on flat lands and gentle slopes (Tanaka, 1986).

It has been expected that boundaries of forest zones are not formed by physiological limits of dominant tree species in each forest zone but by "social limits" imposed by inter-species competition (Imanishi, 1937). Although physical factors of climate and land determine the principal frame for plant life, a plant's distribution depends on the completion of its life cycle in competition with other plants in a plant community. In this sense, biological factors controlling plant survival in a plant community are also important on a micro-scale.

In order to clarify the causes of the elevational distribution pattern of beech along the Pacific Ocean-Sea of Japan gradient, it is necessary to study the structure and dynamics of beech forests in the upper and lower limit areas and effects of environmental changes along the gradient on the competition between beech and other plant species of the warm-temperate and subalpine zones, as well as their physiological requirements and tolerance to environmental factors.

REFERENCES

Environment Agency, 1980. *Important Plant Communities in Japan.* 12 volumes. Environment Agency, Tokyo. (in Japanese).

Imanishi, K., 1937. Elevational zonation of the Japan Alps. *Sangaku,* **32**, 269–364 (in Japanese).

Kira, T., 1948. On the altitudinal arrangement of climatic zones in Japan. *Kanchi-Nogaku,* **2**, 143–173. (in Japanese).

Kira, T., 1976. *Terrestrial Ecosystems – An Introduction* (*Handbook of Ecology,* Vol. 2), Kyoritsu–shuppan, Tokyo, 166 pp.(in Japanese).

Kira, T. and Yoshino, M., 1967. Thermal distribution of Japanese conifers – on the western regions from the Central Japan. In: *Shizen – Ecological Researches* (ed by T. Kira and M. Morishita) , Chuo–koron–sha, Tokyo, 133–161. (in Japanese).

Kitamura, S. and Murata, G., 1971; 1979. *Coloured Illustrations of Woody Plants of Japan*, Hoikusha, Tokyo, 401 pp.(Vol.1); 545 pp.(Vol.2), (in Japanese).

Tanaka, N., 1981. Altitudinal distribution range of beech (*Fagus crenata* Blume) forests along the climatic gradient between the Pacific Ocean and the Japan Sea in Honshu, Japan. *Bull. Biol. Soc. Chiba,* **30**, 67–69. (in Japanese).

Tanaka, N., 1986. Habitat of the *Fagus crenata– Abies mariesii* forest. *Bull. Tokyo Univ. For.,* **75**, 119–140. (in Japanese with English summary).

THE TEMPERATE ELEMENTS OF THE FLORA OF THE NANSEI-SHOTO (THE RYUKYU ISLANDS) AND THE GLOBAL CLIMATIC CHANGE

Hideaki Ohba

ABSTRACT: In the Nansei-Shoto the phytogeography is complicated and presents an interesting problem in studies related to climatic fluctuation in the past and also to climatic changes induced by increasing atmospheric CO_2. In the Nansei-Shoto from sea level to the peaks of mountains at 979 m, varied floras show different origins and affinities. The Islands share the flora of the lowlands with Taiwan, southern China, and Malaysia.

There are also a large number of temperate elements. *Hydrangea* is a representative genus showing complicated speciation in the Islands. There are also various species with extraordinary distribution patterns. *Satakentia*, the only endemic genus in this region, indicates a floristic relationship to the Ogasawara Islands and other Pacific islands. *Hibiscus makinoi* is closely related to *H. pacificus* of the Ogasawara Islands and *H. taiwanensis* of Taiwan. The united ranges resemble the range of *Crossostephium chinense* (Compositae).

Phytogeographically the Nansei-Shoto should be classified into the Southern Asiatic Region with Taiwan and south China, since the area differs considerably from the Sino-Japanese Region.

KEY WORDS: climatic change, flora, Nansei-shoto, Ryukyu Islands, temperate element.

1. INTRODUCTION

The flora of Japan is characterized by high species diversity throughout, even within the physiognomically same vegetation type. The Nansei-Shoto is situated on the very fringe of the continental shelf nearly in the center of the great arc of islands that stretches from the Kuriles and Sakhalin in the north through Japan proper and Taiwan to the Philippines and the Malay Archipelago in the south.

The Nansei-Shoto, consisting of some 150 islands, is situated between Kyushu and Taiwan. The islands are arranged in a chain more than 1,000 km long located between 30° N and 24° N, and geographically divided into five units (groups of islands), Satsunan, Tokara, Amami, Okinawa, and Sakishima (often subdivided into two, Miyako and Yaeyama).

Phytogeographically the Satsunan group, consisting of Yakushima (Yaku Island), Tanegashima, Kuroshima and Kuchino Erabu-jima, are usually excluded from the Nansei-Shoto Islands (Fig. 1). I accept their exclusion in this article. Various

University Museum, University of Tokyo, Hongo 7-3-1, Tokyo 113, Japan.

names have been applied in the literature to the remaining range of islands: Loo Choo, Luchu, Lutchu, Liu Kiu, Liukiu, Ryukyu, Ryukyus, South-West or Nansei-Shoto. The name Nansei-Shoto is accepted as the standard for plant taxonomic databases (Hollis and Brummitt, 1992). The total land area is 3614.57 km^2, and the largest island is Okinawa, 1199 km^2, followed by Amami-Oshima, 712 km^2.

The islands are of considerable elevation notwithstanding their origin from submerged sea-mountains, and the highest elevation is reached on Mt. Mitake on Nakanoshima Island of the Tokara group, which attains an elevation of 978 m. Some small islands are mostly flat and of coral origin. The flora of the Nansei-

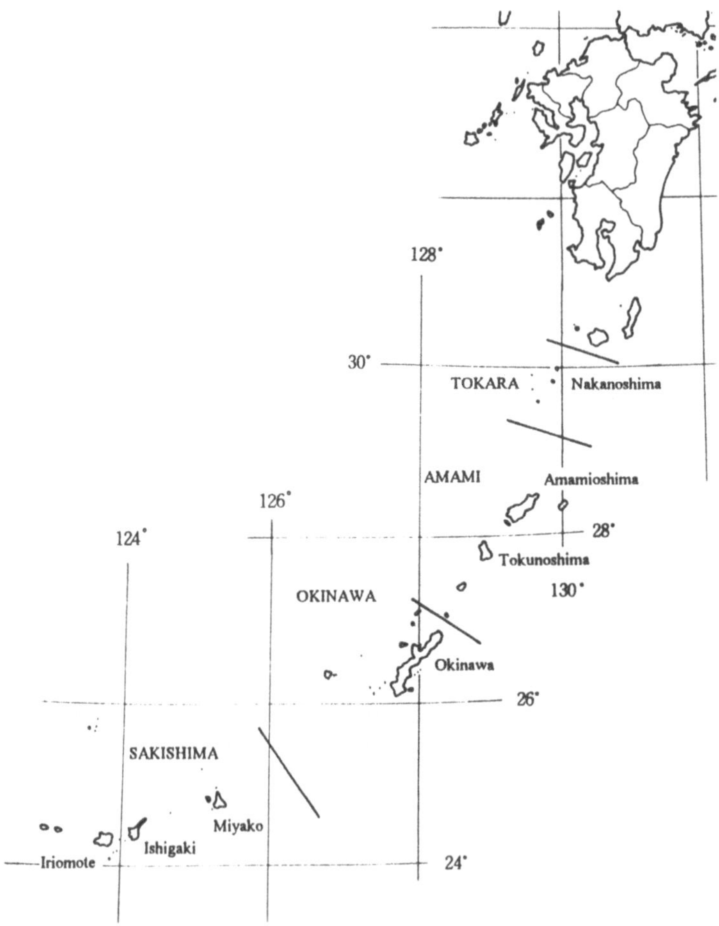

Fig. 1. Geography of the Nansei-Shoto, S. Japan. TOKARA, AMAMI, OKINAWA and SAKISHIMA are names for each group.

Shoto has been studied by many workers (Hooker and Arnott, 1838; Engler, 1884; Forbes and Hemsley, 1886-1905; Makino, 1894-1896; Matsumura, 1897-1898, 1898; Ito and Matsumura, 1898; Warburg, 1899-1900; Kawagoe, 1916; Wilson, 1920; Koidzumi, 1928, 1932, 1933; Sakaguchi, 1924; Masamune, 1934, 1935, 1951-1964; Ito, 1938-1939; Sonohara *et al.*, 1952; Hosokawa, 1958; Hatusima and Amano, 1958; Hatusima, 1960-1962, 1971, 1975; Walker, 1976, and others). However, the taxonomy of the floral elements is still not sufficiently known, and a considerable number of specimens are necessary to make progress on further comparative studies among corresponding taxa, particularly those of Taiwan, China, and the Malaysian regions, as well as those of Japan proper. Notwithstanding such circumstances, Shimabuku (1990) published an enumeration of all the species known from the Nansei-Shoto. This article aims to introduce the characteristics of the flora and to provide discussion regarding their phytogeography and the implication of global climate changes induced by increasing atmospheric CO_2 (Houghton *et al.*, 1990).

2. PHYSICAL FEATURES

In the Nansei-Shoto the temperature exceeds freezing throughout year. The average temperature is 21.3℃ at Nase (Amami-Oshima), 22.4 at Naha (Okinawa), 23.1 at Miyako (Miyakojima), 23.8 at Ishigaki (Ishigakijima). The average temperature in the coldest month is 14.2 at Nase, and 16.0 at Naha. The annual precipitation is above 2000 mm/year on most of the islands. The climate is transitional between marine subtropical and warm temperate.

The vegetation is well developed under such favorable conditions as rather high temperature with ample precipitation and humidity throughout year, and shows a considerable variety of physiognomies and floristic constitutions. The vegetation of the Nansei-Shoto is principally subtropical, composed of evergreen rain forests (Hosokawa, 1958). The ecological northern limit of the subtropical zone may lie on or about the Tokara (Watase's, Aoki's) line or Miyake's line (Ahti *et al.*, 1974).

3. FLORA

The Nansei-Shoto is a natural stepping stone for the migration of floras from north to south and vice versa, and also from China in the west. The location of the Nansei-Shoto in its proximity with Taiwan is unique in phytogeography.

The Nansei-Shoto is thought to be dominated by tropical and subtropical elements. Tropical floristic elements are thought to have invaded from south to north along the island arc. According to Hatusima (1971), 1400 species of seed plants are known from these islands, of which 700 species do not occur in Japan proper north of the Nansei-Shoto. Shimabuku (1990) enumerated 1396 species with 260 pteridophytes, 5 gymnosperms, and 1396 angiosperms (919 dicotyledons and 477 monocotyledons). There are 160 endemic species (including subspecies and varieties) in the Nansei-Shoto (Appendix).

Koidzumi (1932, 1933) pointed out the significance of the Tokara Straits as the floristic boundary between Japan proper and the Nansei-Shoto. Table 1 is a list of the indigenous and endemic species, and also the species at the northern and southern limits of their distribution, together with the size and height of the main islands of the Tokara and Amami groups, and Okinawa and Iriomote islands. Good (1974) classified the islands phytogeographically in the Continental Southeast Asiatic Region of the Paleotropical Kingdom. That is, Good considered the flora of the Nansei-Shoto to be related to Japan proper and eastern China. Masamune (1934) regarded the Nansei-Shoto (which he called the Ryukyu Archipelago and included the Satsunan group) as a single phytogeographical unit and then subdivided into four groups: the Satsunan region (Tanegashima, Yakushima and the adjacent islands), the Amami region (including Amami-Oshima, Tokunoshima), the Okinawa region (Okinawa and adjacent islands), and the Sakishima region. Takhtajan (1969) classified the Nansei-Shoto excluding the Sakishima group into the Eastern Asian Region, which corresponded with Good's Sino-Japanese Region, while he placed the Sakishima group in the Indo-Chinense (Continental South-east Asian) Region. Hatusima (1971) analyzed the distribution of 429 species of the Nansei-Shoto with narrow distribution ranges. Among them, 142 species are distributed in Japan proper and the Nansei-Shoto; 78 are in Japan, the Nansei-Shoto and Taiwan; 85 are in the Nansei-Shoto and Taiwan; 50 are in the Nansei-Shoto, Taiwan and southern China; 45 are in Japan proper, the Nansei-Shoto, Taiwan and southern China; 10 are in Japan, the Nansei-Shoto, Taiwan, and the Philippines; 8 are in the Nansei-Shoto, Taiwan and the Philippines; 6 are in the Nansei-Shoto and Ogasawara; and 5 are in the Nansei-Shoto and southern China.

Table 1 Number of taxa recognized, area and height of each island

Islands	Native taxa				Area	Height
	Indigenous	Endemic	Northern	Southern	(km^2)	(m)
Kuchinoerabushima	381	0	2	1	38	657
Kuchinoshima	467	1	6	6	13	628
Nakanoshima	626	0	7	10	28	979
Gajajima	324	0	1	3	5	497
Hirashima	284	0	1	1	2	250
Suwanosejima	322	0	1	4	22	799
Akusekijima	422	0	9	5	7	584
Kodakarajima	215	0	2	0	1	103
Takarajima	517	0	16	16	6	292
Yokoatejima	91	0	1	1	4	495
Amami-oshima	1016	18	120	14	719	694
Kikaijima	575	1	6	4	56	224
Tokunoshima	759	3	10	9	248	645
Okierabujima	659	0	20	8	95	246
Yoronjima	430	0	3	0	21	97
Okinawa	1243	17	76	33	1199	498
Iriomotejima	1200	8	24	62	289	470

These statistics imply the transitional feature of the flora of the Nansei-Shoto. The physical features of the Nansei-Shoto islands, such as their stepping stone arrangement, climatic conditions, and their geographical location, present fascinating opportunities for biotic invasions and migrations along an island arc.

According to Maekawa (1974) there are clear limit lines where numbers of representative elements suddenly decrease. These lines are between Miyako (the Sakishima group) and Okinawa, and north of Amami-Oshima (i.e., the Tokara Straits). Maekawa pointed out the phytogeographical significance of these two lines and gave evidence from two distinct examples in the sea grasses and mangroves. A large number of tropical genera have their northern limits in the Nansei-Shoto, particularly on Amami-Oshima, Okinawa, and Iriomote (Table 2).

The number of the elements is considered to decrease gradually from southwest to northeastward, however, and the two lines proposed by Maekawa may reveal the phytogeographical significance of large islands with high elevations. Different species show different distribution patterns depending on their preference, tolerance, competition, etc., but they may not survive under temperatures lower than freezing. It is not suitable to set up demarcation lines within the Nansei-Shoto based on the distribution pattern of the tropical elements. In my opinion the Tokara Straits, between the Tokara group and the Satsunan group (Yakushima and Tanegashima) form a significant phytogeographical demarcation line as pointed out by Koidzumi (1932, 1933). This line corresponds to Watase's line, which is proposed chiefly on paleobiogeographical data for vertebrates, and Aoki's line based on entomological studies and also on climatic data. Table 1 shows the islands of the Tokara group to

Table 2 Representative genera which have their northern limit of distributions in the Nansei-Shoto

Islands	Genera
Amami-oshima	*Arthropteris, Ormocarpum, Pongamia, Alysicarpus, Pycnopora, Murrya, Croton, Discocleidon, Macaranga, Maytenus, Abutilon, Heritiera, Helicteres, Thespesia, Barringtonia, Bruguiera, Cerbera, Cordia, Heliotropium, Anisomeles, Thysanospermum, Wendlandia, Apluda, Dactyloctenium, Chryspogon, Aristida, Urochloa, Gahnia, Arenga, Cymodocea, Diplanthera*
Okinawa	*Tapeinidium, Schizaea, Pseudodrynaria, Pipturus, Hiptage, Tristellateia, Ryssopteris, Scolopia, Melanolepis, Cleome, Rhizophora, Lumnitzera, Rhodomyrtus, Evolvulus, Hyptis, Lepidagathis, Atylosia, Heteropogon, Cyrtococcum, Oryzopsis, Epipremnum, Pistia, Acanthophippium*
Iriomote	*Dipteris, Acrostichum, Lomariopsis, Prosaptia, Illigera, Dalbergia, Pterocarpus, Intsia, Mohgania, Uraria, Pithecelobium, Bridelia, Allophylus, Nothapodytes, Kleinhovia, Saurauja, Garcinia, Begonia, Sonneratia, Ecdysanthera, Ochrosia, Gymnema, Lepistemon, Stictocardia, Avicennia, Coleus, Cyrtandra, Hemiboea, Titanotrichum, Hemigraphis, Argostemma, Guettarda, Klnoxia, Freycinetia, Garnotia, Thaumastochloa, Enhalus, Nipa, Rhaphidophora, Anoectochilus, Corymborchis, Spathoglottis, Staurochilus, Sterosandra, Thrixspermum*

be relatively small and the number of the species are, on most islands, less than the half the number on Amami-Oshima. The flora of Amami-Oshima is peculiar. That is, 120 tropical species (142 according to Hatusima (1971) and Maekawa (1974)) not only have their northern limits on Amami-Oshima, but a considerable number of species are endemic to the island. Yakushima (in the Satsunan group), 540 km^2 in size and with the highest elevation, 1935 m on Mt. Miyanouradake, is also phytogeographically significant. More than 1170 species of vascular plants are indigenous to the island, including a large number of endemic species and infraspecific taxa (Yahara *et al.*, 1987).

More than 70 species have their northern limit of distribution and more than 200 species have their southern limit of distribution on Yakushima. Most of these species are temperate elements. The representative species are: *Equisetum arvense, Plagiogyria matsumureana, Torreya nucifera, Cephalotaxus harringtonia, Tsuga sieboldii, Cryptomeria japonica* var. *japonica, Abies firma, Pseudostellaria heterantha, Akebia quinata, Schizophragma hydrangeoides, Parnassia palustris, Aruncus dioicus, Prunus jamasakura, Fragaria nipponica, Pourtheaea villosa, Sorbus commixta, Rhododendron yakusimanum, Tripterospermum japonicum, Viburnum erosum* and *Cardiocrinum cordatum.*

Thus, the Tokara Straits, between the Satsunan group (Yakushima, Tanegashima, etc.) and the Tokara group, is a reasonable phytogeographical demarcation line. Most temperate elements of the flora of Japan proper do not extend beyond the straits. As a global floristic division, the Nansei-Shoto should be classified in the Southern Asiatic Region together with Taiwan and southern China, even though there are 142 species with distributions from the Nansei-Shoto to Japan proper, in addition to species with extraordinary patterns. The flora of the Nansei-Shoto is naturally insular in character, and it is not closely similar to the flora of mainland Asia and the continental features in the flora are not immediately evident. Although the Toraka Straits represent an acceptable demarcation line, no clear demarcation line between the islands in the Nansei-Shoto is recognizable.

4. FLORISTIC RELATION TO THE OGASAWARA ISLANDS

There are some species in common between the Nansei-Shoto and the Ogasawara: *Asplenium cardiophyllum, Celtis boninensis, Osteomeles anthyllidifolia, Photinia wrightiana, Schima wallichii* subsp. *noronhae, Vaccinium wrightii, Corydalis heterocarpa* var. *brachystyla* and *Zanthoxylum beecheyanum.* Among these, *Asplenium ardiophyllum, Celtis boninensis, Corydalis heterocarpa, Photinia wrightiana, Schima wallichii* and *Vaccinium wrightii* are considered to extend from the Asian side to the western Pacific islands.

Hibiscus makinoi Jotani & H.Ohba of the Nansei-Shoto (including the Satsunan group and also southern Kyushu) is very close to *H. pacifica* Jotani & H.Ohba of the Ogasawara Islands (in the Kazan Retto) (Jotani and Ohba, 1984, 1986). Both species are related to *Hibiscus mutabilis,* under cultivation and probably native in southwest China, and also to *H. taiwanensis* S.Y.Hu of Taiwan and *H. indicus* (Burm. f.) Hochr. in having rotate corollas with spreading petals. *Hibiscus pacificus* presumably differentiated from the common ancestor shared with *H. makinoi* and *H. taiwanensis*. The combined range of these three species resembles that of *Crossostephium chinense* (L.) Makino, which shows an extension of the eastern Asian elements to the Ogasawara Islands (Fig. 2). The distribution of *Osteomeles* is curious, since it ranges from south China, Lan Yu Island, Yakushima (south Kyushu) to Hawaii and the Polynesian islands. Except for the common insular species occurring in coastal regions, however, the flora of the Nansei-Shoto differs distinctly in character from that of the Ogasawara, which reflects a much more Pacific origin.

5. *SATAKENTIA*

There are a considerable number of species with extraordinary distribution patterns. *Satakentia liukiuensis* (Hatus.) H.E.Moore is one of the most remarkable examples. *Satakentia* is the sole endemic genus in the Nansei-Shoto, and occurs on Ishigaki and Iriomote in the Sakishima group. The genus is very close to *Clinostigma*, which ranges from the Ogasawara Islands to Samoa, Fiji, New Hebrides, the Solomon Islands and New Ireland and is represented by 13 species (Dransfield, 1981; Uhl and Dransfield, 1987).

Satakentia is different from *Clinostigma* by having a tomentose inflorescence, staminate flowers with well a developed pistillode, and the fruit with more or less apical stigmatic remains. The differences between them indicate that *Satakentia* might have separated from *Clinostigma* along the west margin of the common track of both genera in very recent times.

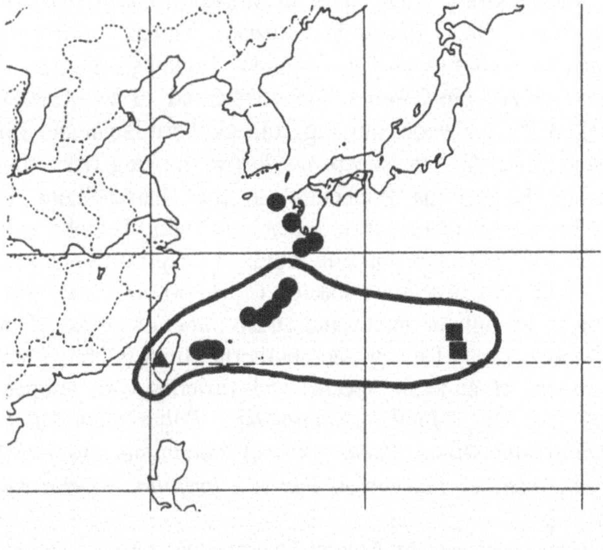

Fig. 2. Distribution of *Hibiscus pacificus* (square), *H. makinoi* (circle), and *H. taiwanensis* (triangle). Line indicates the range of *Crossostephium chinense* (Compositae).

6. THE TEMPERATE ELEMENTS

Table 3 shows the representative temperate genera with ranges limited to the Nansei-Shoto. From middle elevations to the summit there are numerous temperate elements. The species found in the Nansei-Shoto show their greatest affinities with those of Japan proper. It is reasonable to consider that past climatic fluctuation caused migration of temperate elements from Japan proper to the Nansei-Shoto, and these temperate elements may have survived at high elevations on the islands or spread there after adaptation to a subtropical climate.

Table 3 Representative temperate genera which have their southern limits of distributions in the Nansei-Shoto

Island	Genera
Amami-oshima	*Sedum, Cardiandra, Stachyurus, Rhodea, Epipactis*
Okinawa	*Brasenia, Merculiaris, Aucuba, Stachys, Orobanche, Adenophora, Senecio, Inula, Sparganium, Agropogon, Luzula, Vexillabium*
Iriomote	*Trochodendron, Cornus, Hedera*

The representative species of the temperate elements are: *Trochodendron aralioides, Stachyurus praecox, Geranium thunbergii, Cornus kousa, Aucuba japonica, Helwingia japonica, Adenophora triphylla* var. *japonica, Rhodea japonica, Epipactis thunbergii.* Hatusima (1971) pointed out that the populations of the Nansei-Shoto differ slightly from those of Japan proper in *Helwingia, Cornus, Rhodea, Adenophora,* etc. *Elatostema suzukii* Yamaz. (*Pellionia cuneata* T. Suzuki), endemic to Okinawa and *E. oshimensis* (Hatus.) Yamaz. (*P. oshimensis* Hatus.), endemic to Amami-Oshima, are considered to be closely related to *E. radicans* (Siebold & Zucc.) Wedd. ranging from southern Japan and Taiwan to China. *Sedum uniflorum* Hook. & Arn. subsp. *uniflorum*, ranging from southern Kyushu (Osumi Peninsula, Kagoshima Prefecture) to the Nansei-Shoto, has a reduced inflorescence in comparison with subsp. *japonicum* (Miq.) H.Ohba of Japan proper.

In *Cardiandra, C. amamiohsimensis* Koidz. is a primitive species endemic to Amami-Oshima while the derivative species, *C. alternifolia* Siebold & Zucc., subsp. *alternifolia* ranges throughout Japan and subsp. *moellendorffii* (Hance) H.Hara & H.Ohba is in Iriomotejima, Taiwan and southern China (Ohba, 1985). There are also a large number of endemic species and varieties with temperate origins in Amami-Oshima: *Polypodium amamianum, Polystichum obai, Vaccinium amamianum, Oxalis amamiana, Parthenocissus amamiana, Lysimachia liukiuensis, Viola amamiana, Ainsliaea macrocliniodes* var. *latifolia, Shortia rotundifolia* var. *amamiana*, etc.

Some species are endemic to Amami-Oshima and also to Tokunoshima in the same island group: *Rubus amamianus, Zanthoxylum amamiense, Callicarpa oshimensis*, some species of *Heterotropa, Arisaema heterocephalum* subsp.

heterocephalum, etc. Other remarkable cases are *Pieris japonica* var. *koidzumiana, Rhododendron scabrum, Heloniopsis kawanoi, H. leucantha,* and *Disporum sessile* var. *micranthum*, which are distributed on several islands in the Nansei-Shoto.

7.THE SPECIES OF HYDRANGEACEAE

I have investigated speciation in the temperate genera *Deutzia* and *Hydrangea* in the Hydrangeaceae. In the Nansei-Shoto, *Deutzia*, consisting of about 50 species, is represented by two species, *D. naseana* Nakai distributed on Okinawa (var. *amanoi*) and the islands of the Amami group (var. *naseana*), and *D. yaeyamensis* Ohwi, endemic to Iriomotejima in the Sakishima group (Ohba and Akiyama, 1992). These two species are closely related. The chromosome number of *D. yaeyamensis* is 2n=26 while both varieties of *D. naseana* are 2n= 52. The former is diploid and the latter is tetraploid. *Deutzia faberi* (chromosome number is unknown) described from the Kiangsu Hills, Chekiang, east China, and *D. scabra* (2n=130) of Japan proper are close to both *D. naseana* and *D. yaeyamensis*. *Deutzia yaeyamensis* and *D. faberi* are not as vigorous and are probably more primitive than the others. *Deutzia yaeyamensis* hangs from rock cliffs by the sides of water falls while *D. naseana* and *D. scabra* grows in various habitats, such as sunny waste places by

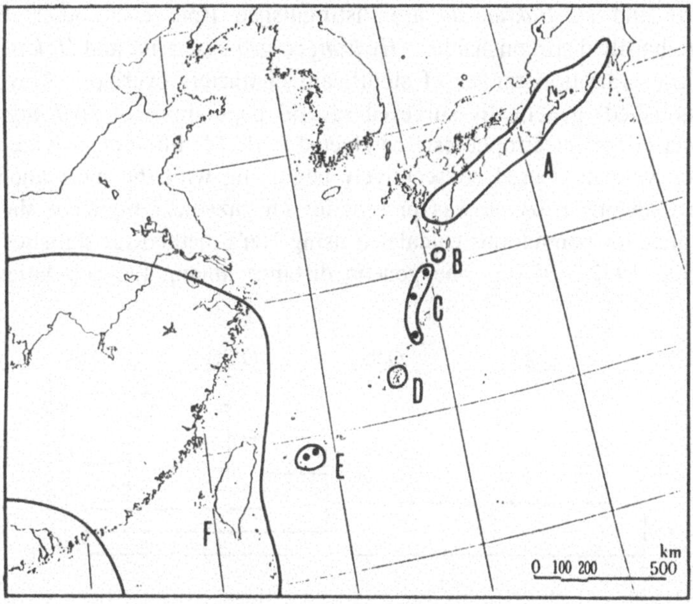

Fig. 3. Distribution of the species of *Hydrangea chinensis* complex. A: *H. scandens.* B & C: *H. kawagoeana* (*Hydrangea* from B often called *H. grosseserrata*). D: *H. liukiuensis.* E: *H. yayeyamensis.* F: *H. chinensis.* *Hydrangea luteovenosa* and *H. hirta* of Japan are excluded.

roadsides. Some species of *Deutzia* are in Taiwan. But the Nansei-Shoto species differ greatly from these of Taiwan, such as *D. taiwanensis* (Maxim.) C.K.Schneider.

In the *Hydrangea chinensis* complex there are several species in the Nansei-Shoto: *H. yayeyamensis* Koidz., *H. liukiuensis* Nakai, *H. kawagoeana* Koidz. and *H. grosseserrata* Engl. (Fig. 3). These are all related to *H. chinensis* Maxim. from Taiwan and southern China and *H. scandens* (Murray) Ser. and *H. luteovenosa* Koidz. from Japan. McClintock (1957) united these and some other taxa into a single species, *H. scandens* with four subspecies, *scandens, liukiuensis, chinensis,* and *kwangtungensis*. *Hydrangea luteovenosa* was reduced to subsp. *liukiuensis*. Hatusima (1971) accepted her concepts with some modifications. *Hydrangea luteovenosa* was separated from subsp. *liukiuensis*. Kitamura (1979) proposed a modified concept of McClintock's *H. scandens* in which he recognized two species in the Nansei-Shoto: *H. liukiuensis* and *H. chinensis* with var. *chinensis* and var. *grosseserrata* including both *H. kawagoeana* and *H. yayeyamensis*. Ohba (1989) treated all four taxa as distinct species. Chromosome numbers of these species (Akiyama, 1989) revealed the presence of polyploidy in the genus. In *H. yayeyamensis* two populations from Iriomotejima were 2n=144 and two from Ishigakijima were 2n=180, while three populations of *H. chinensis* from Taiwan were 2n=36. From the estimated basic chromosome number, *H. yayeyamensis* is octaploid and decaploid while *H. chinensis* is diploid. *Hydrangea grosseserrata* and *H. kawagoeana* with 2n=36 are diploid. Akiyama found two cytoraces in *H. liukiuensis*, diploid with 2n=36 and tetraploid with 2n=72. Morphologically *H. yayeyamensis* and *H. liukiuensis* are distinguished from each other, but their cytotypes are hardly distinguishable. *Hydrangea grosseserrata* and *H. kawagoeana* are quite close: variation ranges of significant characters overlap. Akiyama and Ohba (unpublished) genetically surveyed several populations of *Hydrangea* using horizontal starch gel electrophoresis conducted with 12 different enzyme systems. Isozyme loci were numbered consecutively beginning with the most anodal locus when more than one isozyme was present for an enzyme. Figure 4 shows their genetic distance for populations calculated using Nei's method for statistics on gene diversity (Nei, 1972, 1973). The genetic distance among the populations of *H.*

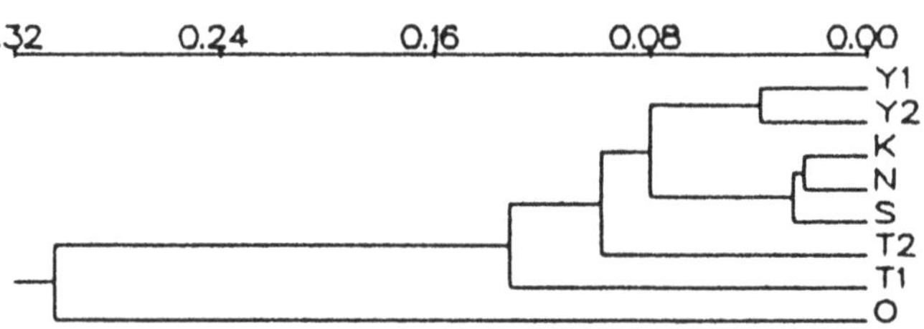

Fig. 4. Cluster diagram based on Nei's (1978) unbiased genetic identity values. Y: Yakushima, K: Kuchinoshima, N: Nakanoshima, S: Suwanosejima, T: Tokunoshima, O: Okinawa. Y1 and Y2, and T1 and T2 were collected from populations in different localities on each island. Except O, all populations belong to *Hydrangea kawagoeana*; O applies to *H. liukiuensis*.

grosseserrata (Y1 and Y2) and *H. kawagoeana* (K, N, S, T1, T2) is less than 0.16, while these populations are separated from the population of *H. liukiuensis* in Okinawa (O) by a value of about 0.32. The genetic distance between *H. grosseserrata* and *H. kawagoeana* is lower than the lowest value normally found between congeneric species (Crawford, 1983). The distance between *H. liukiuensis* and *H. kawagoeana* (incl. *H. grosseserrata*) is considerably large.

Hydrangea yayeyamensis is morphologically similar to *H. chinensis* from Taiwan and southern China, but is unlike *H. liukiuensis* of Okinawa. *Hydrangea liukiuensis* is related to *H. luteovenosa* (2n=36 by Funamoto and Tanaka, 1988) and *H. scandens* of Japan proper. It is interesting that *H. luteovenosa* occurs on Yakushima and is distinguished as var. *yakushimensis* (Masam.) Sugimoto. *Hydrangea liukiuensis*, without ornamental flowers like *H. hirta*, is thought to be more primitive than *H. luteovenosa*, and might be surviving on Okinawa because of genetic multiplication through polyploidy.

To conclude the discussion of Hydrangea in the Nansei-Shoto, there are three species with different geographical distributions, and with complicated historical backgrounds.

8. INFLUENCE OF CLIMATE CHANGE TO THE FLORA

The Intergovernmental Panel on Climate Change (IPCC) has discussed the shifting of vegetational zones and floristic compositions under several scenarios of drastic climate change induced by a global increase in atmospheric CO_2 (Melillo *et al.*, 1990; Ohba, 1994). The effect of climate change due to CO_2 doubling on the natural vegetation of Japan was estimated by Uchijima and Seino (1990) and Tsunekawa *et al.* (1993). Ohba (1994) suggested that small sized plant communities, which are often restricted in their ranges, would have difficulty surviving drastic climate changes because of their narrow range, discontinuous distribution, and special isolation. In relation to communities endemic to Amami-Oshima, the *Quercus amamiana* and the *Arisaema heterocephalum-Castanopsis sieboldii* communities would be seriously threatened. These communities contain a considerable number of endemic species with narrow ranges. These species might disappear because of climate change and deformation or extinction of the communities in which they occur. Furthermore, most temperate elements in the Nansei-Shoto would be seriously threatened. They may have narrowly escaped extinction after migrating to the islands after the last glacial period. At present many endemic species with temperate origins, such as *Vaccinium amamianum* and *Oxalis amamiana*, are difficult to locate because of their rarity. A large number of them may have gone extinct during the last two decades.

To estimate responses to the scenarios for global climate change induced by increasing atmospheric CO_2 the need to make clear the detailed distribution of each species is still urgently needed and taxonomic data about the flora is a vital necessity. These data should also include geography and habitat, life history, and interactions with abiotic factors and with other organisms.

ACKNOWLEDGMENT

This study was funded by a research grant from the Environment Agency, Japan. I am indebted to Dr. David E. Boufford of the Harvard University, Herbaria for the correction of English.

REFERENCES

Ahti, L.H., Ahti, T. and Koponen, T., 1974. A scheme of vegetative zone for Japan and adjacent regions. *Ann. Bot. Fennici,* **11**, 59-88.

Akiyama, S., 1989. Chromosome numbers of the species in *Hydrangea chinensis*-complex in the Ryukyu Islands. *J. Jpn. Bot.,* **64**, 353-360.

Crawford, D.J., 1983. Phylogenetic and systematic inferences from electrophoretic studies. In: *Isozyme in Plant Genetics and Breeding. Part A.* (ed. by S.O. Tanksley and T.L. Orton), 257-287, Elsevier Scientific Publishing Co., Amsterdam.

Dransfield, J., 1981. Palms and Wallace's Line. In: *Wallace's Line and Plate Tectonics.* (ed. by T.C. Whitmore), Oxford University Press, London, 43-56.

Engler, A., 1884. Beitrage zur Flora des sudlichen Japan und der Liu-kiu-Insel. Auf Grund der von Dr. Doderlein und Tachiro gesammelten Pflanzen. *Engler Bot. Jahrb.,* **6**, 49-74.

Forbes, F.B. and Hemsley, W.B., 1886-1905. An enumeration of all plants known from China proper, Formosa, Hainan, Corea, the Luchu Archipelago and the Island of Hongkong, together with their distribution and synonymy. *J. Linn. Soc. Bot.*, **23**, 1-521 (1886-1888); **26**, 1-592 (1889-1902); **36**, 1-686 (1903-1905).

Funamoto, T. and Tanaka, R., 1988. Karyomorphological studies in some taxa of *Hydrangea* from Japan. *Kromosomo II*, **49,** 1583-1594.

Good, R., 1974. *The Geography of the Flowering Plants.* 4 ed. Longmans, London.

Hatusima, S., 1960-1962. Flora of the Amami Islands. *Mem. South. Indust. Sci. Inst. Kagoshima Univ.,* **2**, 21-31 (1960); **3**, 5-41, 113-144 (1962).

Hatusima, S., 1971. *Flora of the Ryukyus.* Okinawa Seibutsu Kyoiku Kenkyukai, Naha.

Hatusima, S., 1975. *Flora of the Ryukyus,* revised edition. Okinawa Seibutsu Kyouiku Kenkyukai, Naha.

Hatusima, S. and Amano, T., 1958. Flora of Okinawa. *Univ. Ryukyu Ext. Ser.,* **15,** 1-192.

Hollis, S. and Brummitt, R.K., 1992. *World Geographical Scheme for Recording Plant Distributions.* Hunt Institute for Botanical Documentation, Pittsburgh.

Hooker, W.J. and Arnott, W.G., 1838. Loo Shoo and Bonin. In: *The Botany of Captain Beechey's Voyage* ([1830]-1841). 258-275, Henry G. Bohn, London.

Hosokawa, T., 1958. On the synchorological and floristic trends and discontinuities in regard to the Japan-Liukiu-Formosa area. *Vegetatio,* **8,** 65-92.

Houghton, J.T., Jenkins, G.J. and Ephraums, J.J., 1990. *Climate Changes. The IPCC Scientific Assessment.* Cambridge University Press, Cambridge.

Ito, H., 1938-1939. Filices Liukiuenses. *Bot. Mag. Tokyo*, **52**, 532-538, 583-590, 642-649 (1938); **53**, 23-28, 68-71 (1939).

Ito, T. and Matsumura, J., 1898. Tentamen florae Lutchuensis. Sectio prima. *J. Coll. Sci. Univ. Tokyo,* **12**, 263-541.

Jotani, T. and Ohba, H., 1984. A wild *Hibiscus, H. makinoi*, sp. nov. from S.W.Japan. *J. Jpn. Bot.*, **59**, 214-222.

Jotani, T. and Ohba, H., 1986. *Hibiscus pacificus* Nakai from the Volcano group of islands. *J. Jpn. Bot.*, **61**, 97-103.

Kawagoe, S., 1916. On the flora of Tokara Islands, especially of Island Naka-no-shima and Island Takaradzima. *Bull. Kagoshima Coll. Agric. Forest,* No. 1, 93-200.

Kitamura, S., 1979. *Hydrangea.* In: *Coloured Illustrations of Woody Plants of Japan.* (ed. by S. Kitamura and G. Murata), Hoikusha Publishing Co. Ltd., Osaka, 2, 111-121.

Koidzumi, G., 1928. *Plantae Novae Amami-Ohsimensis nec non Insularum Adjacentium.* Kagoshimaken Kyoiku Chosakai.

Koidzumi, G., 1932. The Tokara channel as the floristic demarcation line. *Acta Phytotax. Geobot.*, **1**, 182-185.

Koidzumi, G., 1933. Additional notes on the phyto-geographical boundary between Japan, Formosa, and the Liuchiu Islands. *Acta Phytotax. Geobot*, **2**, 206.

Maekawa, F. 1974. Origin and characteristics of Japan's flora. In: *The Flora and Vegetation of Japan.* (ed. by M. Numata), 33-86, Kodansha Limited, Tokyo, and Elsevier Scientific Publishing Co., Amsterdam.

Makino, T., 1894-1896. Mr. Hisashi Kuroiwa's collections of Liu Kiu plants. *Bot. Mag. Tokyo*, **8**, (370)-(373), (411)-(417) (1894); **9**, (6)-(14), (61)-(64), (102)-(106), (255)-(257), (278)-(285), (320)-(328), (379)-(382)(1895); **10**, (9)-(11), (220)-(224), 55-60, 63-68 (1896).

Masamune, G., 1934. On the phytogeography of the Ryukyu Archipelago. *Bull. Biogeograph. Soc. Jpn.*, **5**, 29-86.

Masamune, G., 1935. Review of the flora of the Ryukyu archipelago. *Trans. Natl. Hist. Soc. Formosa,* **25**, 253-254.

Masamune, G., 1951-64. Enumeratio Tracheophytarum Ryukyu Insularum. *Sci. Rep. Kanazawa Univ.*, **1**, 33-54, 167-199 (1951); **2**, 87-114(1953), 59-117 (1954); **3**, 101-182, 253-338 (1955); **4**, 45-134 (1955), 201-280 (1956); **5**, 81-121 (1957); **9**, 119-154 (1964).

Matsumura, J., 1897-1898. Notes on some Liu-kiu plants. *Bot. Mag. Tokyo*, **11**, 69-71, 77-78 (1897); **12**, 1-3, 13-15 (1898).

Matsumura, J., 1898. Notes on Liukiu and Formosan plants. *Bot. Mag. Tokyo*, **12**, 53-55, 61-63, 67-68, 75-76, 83-84, 107-109.

McClintock, E., 1957. A monograph of the genus *Hydrangea.* *Proc. Calif. Acad. Sci.*, **29**, 147-256.

Melillo, J.M., Callaghan, T.V., Woodward, F.I., Salati, E. and Sinha, S.K., 1990. Effect on ecosystems. In: *Climate Changes. The IPCC Scientific Assessment,* (ed. by J.T. Houghton, G.J. Jenkins and J.J. Ephraums), 282-310, Cambridge University Press.

Nei, M., 1972. Genetic distance between populations. *Am. J. Bot.*, **63**, 2448-2453.

Nei, M., 1973. Analysis of gene diversity in subdivided population. *Proc. Natl. Acad. Sci. U. S. A.*, **70**, 3321-3323.

Nei, M., 1978. Estimation of average heterozygosity and genetic distance from a small number of individuals. *Genetics*, **89**, 583-590.

Ohba, H., 1985. A systematic revision of the genus *Cardiandra* (Saxifragaceae-Hydrangeoideae). *J. Jpn. Bot.*, **60**, 139-147, 161-171.

Ohba, H., 1989. *Hydrangea.* In: *Wild Flowers of Japan. Woody Plants.* (ed. by Y. Satake, H. Hara, S. Watari and T. Tominari), Heibonsha Publishing Co. Ltd., Tokyo, 1, 166-172.

Ohba, H., 1994. The flora of Japan and the implication of global climatic change. *J. Plant Res.*, **107**, 85-89.

Ohba, H. and Akiyama, S., 1992. A taxonomic revision of *Deutzia* (Saxifragaceae, s.l.) in the Ryukyu Islands, S Japan. *J. Jpn. Bot.*, **67**, 154-165.

Sakaguchi, S., 1924. *General Index to the Flora of Okinawa.* Ishizuka Shoten, Naha.

Shimabuku, K., 1990. *Check List Vascular Flora of the Ryukyu Islands.* Hirugi-sha, Naha.

Sonohara, S., Tawada, S. and Amano, T., 1952. *Flora of Okinawa.* US Civil Administration of the Ryukyu Islands, Naha.

Takhtajan, A., 1969. *Flowering Plants, Origin and Dispersal.* Oliver and Boyd, Edinburgh. (English translation from the Russian original by C. Jeffrey).

Tsunekawa, A., Ikeguchi, H., Kainuma, T., Tobe, K. and Omasa, K., 1993. Effects of climate change on the vegetation distribution in Japan. *J. Agr. Meteorol.*, **48**, 867-870. and also in this volume.

Uchijima, Z. and Seino, H., 1990. Probable effects of CO_2-induced climate change on natural vegetation of Japan. *Report of Studies of Evaluation of Probable Effects of CO_2-induced Climate Change on Natural Ecosystem and Agro-sociological System in Japan*, 49-74. (in Japanese).

Uhl, N. W. and Dransfield, J., 1987. *Genera Palmarum.* Allen Press, Lawrence.

Walker, E., 1976. *Flora of Okinawa and the Southern Ryukyu Islands.* Smithsonian Institution Press., Washington, D.C.

Warburg, O., 1899-1900. *Monsunia.* Verlag von Wilhelm Engelmann, Leipzig.

Wilson, E.H., 1920. The Liukiu Islands and their ligneous vegetation. *J. Arnold Arb.*, **1**, 171-186.

Yahara, T., Ohba, H., Murata, J. and Iwatsuki, K., 1987. Taxonomic review of vascular plants endemic to Yakushima Island, Japan. *J. Fac. Sci., Univ. Tokyo, Sect. III*, **14**, 69-119.

APPENDIX

List of the Endemic Species and Varieties of the Spermatophytes in the Nansei-Shoto

Gymnospermae

Pinaceae

Pinus lutchuensis Mayr
Tokara Group

Angiospermae - Dicotyledones

Fagaceae

Castanopsis cuspidata (Thunb. ex Murray) Schottky var. *lutchuensis* (Koidz.) Masam.
South from Amami-oshima
Quercus miyagii Koidz.
Amami-oshima, Tokunoshima, Okierabujima, Okinawa, Kumejima, Ishigakijima, Iriomotejima

Ulmaceae

Celtis biondii Pamp. var. *insularis* Hatus.
Mikakojima, Taramajima

Urticaceae

Boehmeria yaeyamensis Hatus.
Ishigakijima, Yonakunijima, Yoronjima
Elatostema oshimensis (Hatus.) Yamaz.
Amami-oshima
Elatostema suzukii Yamaz.
Okinawa
Elatostema yonakuniense Hatus.
Yonakunijima

Balanophoraceae

Balanophora kuroiwai Makino.
Okinawa, Miyako-jima, Ishigakijima, Iriomotejima, Yonakunijima

Portulacaceae

Portulaca okinawensis Walker & Tawada
Amami-oshima, Tokunoshima, Okinawa, Kumejima, Awaguni-jima, Tokashikijima, etc.

Magnoliaceae

Illicium anisatum L. var. *tashiroi* (Maxim.) Walker
Ishigakijima, Iriomotejima

Lauraceae

Cinnamomum doederleinii Engl. var. *doederleinii*
Amami-oshima, Tokunoshima, Okierabujima, Okinawa Group, Ishigakijima, Iriomotejima
Cinnamomum doederleinii Engl. var. *pseudodaphoides* Hatus.
Amami-oshima, Keramajima
Lindera communis var. *okinawensis* Hatus.
Okinawa, Ishigakijima
Neolitsea sericea (Bl.) Koidz. var. *argentea* Hatus.
Daitojima

Ranunculaceae

Clematis grata Wall. var. *ryukyuensis* Tamura
Throughout
Ranunculus extorris Hance var. *lutchuensis* (Nakai) Tamura
Tokunoshima, Okierabujima, Okinawa, Iejima

Aristolochiaceae

Aristolochia liukiuensis Hatus.
Throughout
Hetrotropa celsa (Hatus.)
Amami-oshima
Hetrotropa dissita (Hatus.)
Ishigakijima, Iriomotejima
Hetrotropa fudsinoi (T.Ito)
Amami-oshima
Hetrotropa glasina (Hatus.)
Iriomotejima
Hetrotropa gusuk (Yamahata)
Amami-oshima
Hetrotropa hatsushimae (Hatus.)
Tokunoshima

Hetrotropa lutchuensis (T.Ito ex Koidz.)
Amami-oshima
Hetrotropa monodoriflora (Hatus. & Yamahata)
Iriomotejima
Hetrotropa okinawensis (Hatus.)
Okinawa
Hetrotropa pellucidum (Hatus. & Yamahata)
Amami-oshima
Hetrotropa senkakuinsularis (Hatus., *nom. nud.*)
Uotsurijima
Hetrotropa similis (Hatus.)
Tokunoshima
Hetrotropa trinacriformis (Yamahata)
Amami-oshima, Kakeroma-jima, Ukeshima
Hetrotropa yaeyamense (Hatus.)
Iriomotejima

Theaceae

Adinandra ryukyuensis Masam.
Okinawa
Adinandra yaeyamensis Ohwi
Ishigakijima, Iriomotejima
Camellia lutchuensis T.Ito
Okierabujima, Okinawa, Kumejima, Ishigakijima, Iriomotejima
Tutcheria virgata (Koidz.) Nakai
Okierabujima, Okinawa, Kumejima, Ishigakijima, Iriomotejima
Eurya osimensis Masam. var. *osimensis*
Amami-oshima, Tokunoshima, Okinawa
Eurya osimensis Masam. var. *kanehirae* (Hatus.) Hatus.
Amami-oshima, Tokunoshima, Okinawa, Ishigakijima, Iriomotejima
Eurya sakishimensis Hatus.
Ishigakijima, Iriomotejima
Eurya yaeyamensis Masam.
Ishigakijima, Iriomotejima
Eurya zigzag Masam.
Okinawa

Hypericaceae

Hypericum senkakuinsulae Hatus.
Uotsurijima

Saxifragaceae

Cardiandra amamioshimensis Koidz.
Amami-oshima
Deutzia naseana Nakai var. *naseana*
Kikaijima, Amami-oshima, Tokunoshima, Kakeromajima, Okierabujima
Deutzia naseana Nakai var. *amanoi* (Hatus.) H.Ohba
Okinawa
Deutzia yaeyamensis Ohwi
Iriomotejima
Hydrangea liukiuensis Nakai
Okinawa
Hydrangea yayeyamensis Koidz.
Ishigakijima, Iriomotejima

Rosaceae

Rubus amamianus Hatus. & Ohwi *amamianus*
Amami-oshima, Tokunoshima
Rubus amamianus Hatus. & Ohwi var. *minor* Hatus.
Amami-oshima
Rubus tawadanus Koidz.
Tokunoshima, Okinawa
Rubus uchinensis Koidz.
Okinawa

Oxalidaceae

Oxalis amamiana Hatus.
Amami-oshima

Euphorbiaceae

Alchornea liukiuensis Hayata
Amami-oshima, Okierabujima, Yoronjima, Okinawa, Iheyajima, Miyakojima, Ishigakijima, Iriomotejima
Excoecaria formosana Hayata var. *daitoinsularis* (Hatus.) Hatus.
Minami-daitojima

Rutaceae

Zanthoxylum amamiense Ohwi
Kikaijima, Amami-oshima, Tokunoshima, Okierabujima
Zanthoxylum schinifolium Siebold & Zucc. var. *okinawense* (Nakai) Hatus.
Amami-oshima, Tokunoshima, Okinawa

Aceraceae
Acer oblongum Wall. subsp. *itoanum* (Hayata) Hatus.
Okierabujima, Yoronjima, Okinawa, Tokashikijima, Kumejima

Aquifoliaceae
Ilex dimorphophylla Koidz.
Amami-oshima
Ilex integra Thunb. var. *brachypoda* (S.Y. Hu) Hatus.
Amami-oshima, Okinawa
Ilex wauburgii Loesn.
Amami-oshima, Tokunoshima, Okierabujima, Okinawa, Ishigakijima, Iriomotejima

Rhamnaceae
Rhamnella inaequilatera Ohwi
South from Amami-oshima
Rhamnus calicicola Hatus.
Okinawa
Rhamnus kanagusuki Makino
Okinawa

Vitaceae
Parthenocissus amamiana Hatus.
Amami-oshima, Tokunoshima, Okinawa, ?Ishigakijima

Violaceae
Viola amamiana Hatus.
Amami-oshima
Viola stoloniflora Yokota & Higa
Okinawa
Viola tashiroi Makino
Ishigakijima, Iriomotejima
Viola utchinensis Koidz. var. *utchiensis*
Okinawa

Cucurbitaceae
Melothria liukiuensis Nakai
Throughout
Trichosanthes ishigakiensis Walker
Miyakojima, Ishigakijima, Iriomotejima
Trichosanthes miyagii Hayata
Amami-oshima, Tokunoshima, Okinawa

Melastomataceae
Bredia okinawensis (Matsum.) H.L.Li
Okinawa
Bredia yaeyamensis (Matsum.) H.L.Li
Ishigakijima, Iriomotejima

Cornaceae
Helwingia japonica (Murray) F.G.Dietrich var. *liukiuensis* (Hatus.) H.Hara & Kurosawa
Amami-oshima, Tokunoshima, Okinawa, Iheyajima

Araliaceae
Fatsia japonica (Thunb.) Decne. & Planch. var. *liukiuensis* Hatus. (*nom. nud.*)
Kikaijima, Amami-oshima, Tokunoshima, Okinawa, Iheyajima, Daitojima

Diapensiaceae
Shortia rotundifolia (Maxim.) Maxim. var. *amamiana Ohwi*
Amami-oshima

Ericaceae
Pieris japonica var. *koidzumiana* (Ohwi) Masam.
Amami-oshima, Okinawa
Rhododendron amanoi Ohwi
Kumejima, Ishigakijima, Iriomotejima
Rhododendron scabrum D. Don
Amami-oshima, Okierabujima, Okinawa, Keramajima
Rhododendron simsii Planch. var. *tawadae* (Ohwi) Hatus.
Uotsurijima
Vaccinium amamianum Hatus.
Amami-oshima

Primulaceae
Lysimachia liukiuensis Hatus.
Amami-oshima

Oleaceae
Ligustrum liukiuense Koidz.
Amami-oshima, Tokunoshima, Okinawa, Ishigakijima, Iriomotejima
Ligustrum tamakii Hatus.
Tonakijima, Yonakunijima

Osmanthus okinawensis Hatus. (*nom. nud.*)
Okinawa

Loganiaceae
Gardneria liukiuensis Hatus.
Okierabujima, Okinawa, Tonakijima, Kikaijima

Gentianaceae
Swertia kuroiwai Makino
Okinawa, Kumejima
Swertia makinoana F.Maek.
Ishigakijima, Iriomotejima

Asclepiadaceae
Cynanchum matsumurae Yamaz.
Amami-oshima, Okierabujima, Yoronjima, Okinawa Group
Cynanchum wilfordii (Maxim.) Hemsl. var. *amamianum* Hatus.
Amami-oshima
Cynanchum yonakuniense Hatus.
Yanakunijima
Tylophora tanakae Maxim. var. *glabrescens* Yamaz.
Daitojima, Miyakojima, Yonakunijima, Uotsurijima

Rubiaceae
Damnacanthus biflorus (Rehder) Masam.
Amami-oshima, Tokunoshima, Okinawa
Damnacanthus indicus K.F.Gaertn. var. *parvispinus* Koidz.
Amami-oshima, Okinawa
Damnacanthus okinawensis Hatus.
Okinawa
Hedyotis luxurians (Hatus.) Hatus.
Daito
Hedyotis tenelliflora Blume var. *longipes* Hatus.
Iriomotejima
Mussaenda parviflora Miq. var. *yaeyamensis* (Masam.) Yamaz.
Miyakojima, Ishigakijima, Iriomotejima
Ophiorrhiza japonica Blume var. *amamiana* Hatus.
Amami-oshima
Tarenna gyokushinkwa Ohwi var. *yaeyamensis* (Masam.) Hatus.
Ishigakijima, Iriomotejima

Verbenaceae
Callicarpa oshimensis Hayata
Amami-oshima, Tokunoshima, Okinawa, Ishigakijima, Yonakunijima

Labiatae
Salvia pygmaea Matsum. var. *pygmaea*
Amami-oshima, Tokunoshima, Okinawa, Kumejima, Ishigakijima, Iriomotejima
Salvia pygmaea Matsum. var. *simplicior* Hatus.
Amami-oshima, Tokunoshima
Scutellaria kikaiinsularis Hatus.
Kikaijima
Suzukia luchuensis Kudo
Okinawa, Kumejimai, Yonakunijima

Scrophulariaceae
Veronicastrum liukiuense (Ohwi) Yamaz.
Amami-oshima, Kikaigashima, Okinawa

Acanthaceae
Strobilanthes glandulifera Hatus.
Okinawa, Ishigakijima, Iriomotejima
Strobilanthes tashiroi Hayata
Takarajima, Kikaijima, Okierabujima, Okinawa, Iheyajima, Kumejima

Gesneriaceae
Rhynchotechum discolor (Maxim.) B.L.Burtt var. *incisum* (Ohwi) Walker
Okinawa

Caprifoliaceae
Lonicera japonica Thunb. var. *miyagusukiana* Makino
Tokunoshima, Okinawa, Miyakojima, Iriomotejima, Yonakunijima
Viburnum tashiroi Nakai
Amami-oshima, Tokunoshima

Campanulaceae
Lobelia loochooensis Koidz.
Amami-oshima, Okinawa, Kumejima

Compositae

Ainsliaea latifolia Kitam.
Amami-oshima
Ainsliaea macrolinidioides Hayata var. *okinawensis* (Hayata) Kitam.
Tokara Group, Amami-oshima, Tokunoshima, Okierabujima, Okinawa, Kumejima, Ishigakijima, Iriomotejima, Senkaku
Ainsliaea oblonga Koidz.
Okinawa
Aster asa-grayi Makino
Amami-oshima, Okierabujima, Okinawa
Aster miyagii Koidz.
Amami-oshima, Tokunoshima, Okinawa
Aster taiwanensis Kitam. var. *lucens* (Kitam.) Kitam.
Ishigakijima, Iriomotejima, Uotsurijima
Aster walkeri Kitam.
Yonakunijima
Cirsium brevicaule A. Gray var. *brevicaule*
Throughout
Cirsium brevicaule A. Gray var. *irumtiense* Kitam.
Kumejima, Miyakojima, Ishigakijima, Iriomotejima, Yonakunijima
Cirsium brevicaule A. Gray var. *oshimense* Kitam.
Kikaijima, Amami-oshima, Tokunoshima
Eupatorium tawadae Kitam.
Okinawa
Farfugium japonicum (L.) Kitam. var. *luchuense* (Masam.) Kitam.
Amami-oshima, Okinawa, Iriomotejima
Ixeris debilis (Thunb.) A. Gray subsp. *liukiuensis* Kitam.
Amami-oshima, Okierabujima, Okinawa, Miyakojima, Tarajima
Ixeris nakazonei Kitam.
Okinawa, Kumejima, Miyakojima, Ishigakijima
Lagenophora mikadoi Koidz.
Amami-oshima, Okinawa, Iriomotejima
Solidago virgaurea L. var. *insularis* (Kitam.) H.Hara
South of Amami-oshima

Angiospermae - Monocotyledones

Liliaceae

Allium amamianum Tawada
Amami-oshima
Heloniopsis kawanoi (Koidz.) Honda
Amami-oshima, Okinawa, Ishigakijima, Iriomotejima
Heloniopsis leucantha (Koidz.) Hatus.
Tokunoshima, Okinawa, Ishigakijima, Iriomotejima
Lilium japonicum Houtt. var. *alexandrae* Baker
Amami-oshima
Liriope tawadae Ohwi
Okinawa
Rohdea japonica (Thunb.) Roth & Kunth var. *latifolia* Hatus.
Amami-oshima

Burmanniaceae
Saionia shinzatoi Hatus.
Okinawa

Eriocaulaceae

Eriocaulon lutchuense Koidz.
Amami-oshima, Okinawa

Gramineae

Aristida takeoi Ohwi
Amami-oshima, Tokunoshima, Okinawa, Izena
Chikusichloa brachyanthera Ohwi
Iriomotejima
Eccoilopus cotulifer (Murray) A.Camus var. *densiflorum* Ohwi
South of Amami-oshima
Isachne lutchuensis Hatus.
Iriomotejima
Isachne subglobosa Hatus. & T.Koyama
Ishigakijima, Iriomotejima, Yonagunijima
Ischaemum aureum (Hook. & Arn.) Hack.
Throughout excluding Daitojima
Saccarum kanashiroi Ohwi
Okinawa

Palmae

Satakentia liukiuensis (Hatus.) H.E.Moore
Ishigakijima, Iriomotejima

Araceae

Alocasia okinawensis Tawada
Okinawa, Iriomotejima, Hateruma
Arisaema heterocephalum Koidz. subsp. *heterocephalum*
Amami-oshima, Tokunoshima
Arisaema heterocephalum Koidz. subsp. *majus* (Seriz.) J.Murata
Amami-oshima
Arisaema heterocephalum Koidz. subsp. *okinawense* Ohashi & J.Murata
Okinawa
Arisaema kaeashimae Seriz.
Tokunoshima

Cyperaceae

Carex collifera Ohwi
Okinawa, Ishigakijima
Carex kimurae Ohwi & T. Koyama
Tokunoshima
Carex maculata Boott var. *tetsuoi* (Ohwi) T.Koyama
Okinawa
Carex sakonis T.Koyama
Amami-oshima, Tokunoshima, Iheyajima
Carex tamakii T.Koyama
Okinawa
Carex tenuior T.Koyama & H.Chung
Amami-oshima
Eleocharis wichurae Boeck. var. *liukiuensis* (Makino) Ohwi
Okinawa
Fimbristylis leptoclada Benth. var. *takamineana* (Ohwi) T.Koyama
Ishigakijima

Musaceae

Musa liukiuensis (Matsum.) Makino ex Kuroiwa
Throughout

Zingiberaceae

Alpinia koidzumiana Kitam.
Iriomotejima

Orchidaceae

Calanthe amamiana Fukuyama
Amami-oshima
Calanthe discolor Lindl. var. *kanashiroi* Fukuyama
Okinawa
Calanthe tokunoshimensis Hatus. & Ida
Tokunoshima
Dendrobium okinawensis Hatus. & Ida
Okinawa
Gastrodia shimizuana Tuyama
Iriomotejima
Lecanorchis brachycarpa Ohwi
Okinawa, Iriomotejima
Lecanorchis flavicans Fukuyama
Ishigakijima, Iriomotejima
Liparis uchiyamae Schlechter
Amami-oshima
Platanthera sonoharai Masam.
Okinawa

Concluding Remarks

East Asia has the following peculiarities in the natural environment, which characterizes the plant production and ecosystems in macro-scale.

First, East Asia is located between the Eurasian continent, the largest landmass of the world, and the North Pacific Ocean, the largest sea surface in the Northern Hemisphere. Between the Eurasian continent and Japanese Islands, there is Japan Sea, which acts as vapor source region in winter, resulting in anomalously heavy snow on the Japan Sea side of the Islands. Secondly, East Asia is locating at down stream of the Tibetan Plateau, the highest mountain mass, as seen from the tropospheric westerlies, which develop to the strongest in the world in winter. Sharp wind effects associated with snow are important factors for the mountain ecosystems on the alpine and sub-alpine zones. Thirdly, Southeast and East Asia is composed of so many islands and peninsulas, having long coastal line and topographical complexity. This situation causes varieties of plant and animal lives and ecosystems in regional scale as well as in local scale. Material and nutrient transport from the mountain areas to the coastal zones, which provide important fields for human life, is significant. The coastal zone is needless to say the most sensitive for the sea level change. Fourthly, monsoons in summer and winter develop clearly, resulting in seasonal difference in precipitation between the wet and the dry seasons. In addition, tropical cyclones and severe storms with disastrous wind and rain hit frequently the coasts of the regions. All these conditions are to be considered as background of the plant life and ecosystems in East Asia, together with the changing global climate.

Part I of this monograph deals with the climatic change scenarios and the modeling and prediction of potential effects on plant production, vegetation distribution, phenological events, and lake ecosystems. Important points are as follows: Climatic changes under the influence of doubled CO_2 were estimated for Monsoon Asia according to three GCMs. It was indicated that annual temperature increase will be fairly large in the regions of higher latitudes with arid and cool climates. Precipitation change, on the other hand, will be uncertain. An attempt was made to calculate the biomass production of natural vegetation and its probable change affected by doubled CO_2 in Monsoon Asia. It was shown that, if natural vegetation could adapt to rapid shift of climatically suitable habitats for each species through the fast migration of plant species, the total net production will increase by about 10%. It was indicated that the total net production of Indonesia accounts for about 88% of its potential total net production now, which is higher than those values for Japan (75%) and for the world (60 %). But it should decrease rapidly in near future because of population growth and accordingly rapid economic development of Indonesia.

The climatic changes will cause the considerable changes in the vertical distribution of vegetation, permafrost zone, and net plant productivity on Tibetan Plateau. It was expected that the increment in annual mean temperature by 4℃ and annual precipitation by 10% under doubled CO_2 condition would accelerate the speed of desertification which is spreading on this area mainly due to the intensive mass land use at present.

Although temperature gives a negligibly small influence on the CO_2 effect on biomass production of rice plants, it affected significantly the grain yield through spikelet fertility. Higher temperature results in damage on rice spikelets at the moment of flowering and, accordingly, makes them sterile. The results of simulation studies on the effects of doubled CO_2 climates showed that 1.6℃ difference in the spikelet sensitivity to high temperature damage gives significantly large effects on rice yield in Central Japan on the Pacific side and Southwestern Japan. Even the high temperature tolerant genotype, rice yield in these region will decrease by 0-10% compared with that at present, while those in North Japan will increase by 10-25% depending on climate scenarios. The model also predicted that the year-to-year variability of rice yield will increase in the warmer part of Japan as described above under the condition of doubled CO_2, but stabilize in the cooler climate region in North Japan. These tendencies are very important to predict the further rice production in the lower latitudes as well as in the higher latitudes under the global warming.

Shift of vegetation is one of the remarked topics on the effects of global warming. A model study came to a conclusion that the percentage of grid-cells in all of Japan showing different predicted vegetation classification between the present condition and the future warming conditions is 23% for a 1℃ increase, 44% for a 2℃ increase, and 62% for a 3℃ increase. Similar approach to the vegetation shift in China revealed that (i) the conifer forest in Northeast China at the present time may be deprived of its habitat within the boundary of China, (ii) the broadleaved forest in East China may shift northward by 3 degrees of latitude under the 2℃ warming of annual air temperature with 20% increase of annual precipitation, and (iii) desert region in West China may expand and steppe and savanna may decrease especially in the case of 4℃ increase with 20% increase of annual precipitation. The local differences should be studied in the next step, because they depend upon the various vegetation types from region to region and the climatic conditions such as the middle latitudes, but also in the lower latitudes comparing the conditions of rainy season and dry season.

Relationships between phenological dates of 10 species and the climatic elements, based on the observed data at 102 stations in Japan for the period 1953-1990, were studied and simple and multiple regression equations were obtained. On the other hand, making use of the 30-year 1 km^2 temperature-climate mesh-files, phenological dates for each mesh were estimated for 1-3℃ rise of mean temperature by applying the regression equations. Fourteen color plates were presented, which indicate a striking shift of phenological events of various plants particularly in spring and autumn. Another study on the plant phenology in Japan, Korea and China made clear also shifting of the dates, effects of urbanization, difference between El Niño event and La Niña event years, etc. in Monsoon Asia. Also, the difference in sugar cane production between spring and summer plantings in El Niño event and La Niña event years was demonstrated in an island located in the southern limit of Japan.

By using a newly developed combined water temperature-ecological model for small lakes, it was indicated that 2-4℃ air temperature warming will (i) increase thermal stratification in summer, which increases nutrient concentrations in bottom water due to nutrient release from bottom sediment, (ii) increase the concentration of phytoplankton at the beginning of the autumn circulation period and (iii) change the composition of phytoplankton species.

Such empirical approach and numerical experiment by assessing the changes of phenological events and ecosystems are needed for the other regions under the various climatic zones and for the other scales of terrestrial or water areas such as lakes. It is also needed to develop similar approach to assess the other phenological events in four seasons respectively in the middle and higher latitudes and in dry and wet seasons in the lower latitudes.

Part II of the present monograph is dedicated mainly to the researches on mountain vegetation in Japan. Among the topics, snow patches, creeping pine development, subalpine forests, and beech forests, have been dealt with in detail. It is indicated through the analysis of fossil peat soils that "Medieval Warm Period" was prevailed in subalpine meadows in Northeast Japan. When the warming will exceed 2℃, snow patches and alpine meadows will shrink considerably and when it will be accompanied by decreasing snowfall and snow accumulation, the shrinkage will be accelerated and many meadows will disappear.

From the standpoint of ecosystem, it is suggested that the subalpine conifers expand into the alpine regions under the influence of global warming, but it will hardly succeed to alternate the vegetation zone, where well-developed creeping pine (*Pinus pumila*) scrub has already occupied. Another study revealed relationships between the effective accumulated temperature and the phenological stages of *Faulia crista-galli* are shown in terms of locations on the slope, which vary mainly with the distance from the snow patch. Similarly, in the different scale, the distribution area of beech (*Fagus crenata*) on the Japan Sea side of Honshu will be shifted upwards on the mountain slopes, but it will be affected by snow cover status. So that, indirect effect of global warming should be considered for such vegetation dynamics.

It has been suggested on the basis of a pytogeographical study on Nansei-Shoto (Ryukyu Islands) that the small sized plant communities, which are often restricted in their ranges, will be suffered more from drastic climatic change, because of their narrow range, discontinuous distribution, and spatial isolation. This is of special importance to consider the impact of global warming on the vegetation not only on the mountain areas but also on the islands like Nansei-Shoto.

At last, we hope that the results presented in this volume will contribute to built a mile stone on the way of global change studies.

Masatoshi Yoshino
Zenbei Uchijima
Kenji Omasa
Keiko Kai
Hiroshi Taoda

Index

The manufacturer's authorised representative in the EU is Springer Nature Customer Service Centre GmbH, Europaplatz 3, 69115 Heidelberg, Germany. If you have any concerns regarding our products, please contact ProductSafety@springernature.com

Printed and bound by CPI Group (UK) Ltd, Croydon, CR0 4YY
25/09/2026
02222891-0002